Theology, Evolution and the Mind

Theology, Evolution and the Mind

Edited by

Neil Spurway

Theology, Evolution and the Mind, Edited by Neil Spurway

This book first published 2009. The present binding first published 2009.

Cambridge Scholars Publishing

12 Back Chapman Street, Newcastle upon Tyne, NE6 2XX, UK

British Library Cataloguing in Publication Data
A catalogue record for this book is available from the British Library

Copyright © 2009 by Neil Spurway and contributors

All rights for this book reserved. No part of this book may be reproduced, stored in a retrieval system, or transmitted, in any form or by any means, electronic, mechanical, photocopying, recording or otherwise, without the prior permission of the copyright owner.

ISBN (10): 1-4438-1369-9, ISBN (13): 978-1-4438-1369-3

TABLE OF CONTENTS

The Science and Religion Forum ix

Introduction 1
Neil Spurway

PART 1

Chapter One 10
The Prehistory of the Religious Mind
Steven Mithen

Chapter Two 31
Whence Comes Religion? Mithen on Prehistory and Mind
Celia Deane-Drummond

Chapter Three 42
Diverging Paths in the Study of the Evolution of Religion
Lluis Oviedo

Chapter Four 53
Darwin's Gifts to Theology
Fraser Watts

Chapter Five 68
Timeo Darwinos et Dona Ferentes: A Response to Fraser Watts
Anthony Freeman

Chapter Six 74
What can Evolved Minds Know of God? Reconsidering Theology
in the Light of Evolutionary Epistemology
Neil Spurway

Chapter Seven 93
Reply to Professor Spurway
Derek Stanesby

Chapter Eight 104
Are We Ghosts or Machines?
Roger Trigg

Chapter Nine 119
The Emergent Threefold Self: A Response to Roger Trigg
Anne L.C. Runehov

Chapter Ten 130
Unfolding Conversation: A theological Reflection on the Evolution
of the Brain/Mind in *Homo sapiens*
Jeremy Law

Chapter Eleven 155
Response to Jeremy Law
Roger Knight

Chapter Twelve 161
Theology, Evolution and the Mind
Roger Paul

PART 2

Chapter Thirteen 178
Cosmic Conversation: the Evolving Dialogue in Mathematics
between Mind and Reality
Gavin Hitchcock

Chapter Fourteen 189
Are Attitudes to Life Changing? –The Emergence of New Moral Intuitions
R.I. Vane-Wright

Chapter Fifteen 199
Nolitional Freedom and the Neurobiology of Sin
Ron Choong

Chapter Sixteen 206
The Healing Instinct: Functionality, Integrity and Relationship –
Holistic Principles in Evolution
Jeremy Swayne

Chapter Seventeen 213
The Mind in Theological and Scientific Perspective
Sjoerd L. Bonting

Index 238

THE SCIENCE AND RELIGION FORUM: SEEKING BOTH INTELLIGIBILITY AND MEANING

Growing out of informal discussion meetings which began in 1972, around the key figure of Revd Dr Arthur Peacocke, the Science And Religion Forum was formally inaugurated in 1975. Its stated purpose was to enable and encourage further discussion of the issues which arise in the interaction between scientific understanding and religious thought. These issues, together with the social and ethical decisions demanded by scientific and technological advances, have remained the subject of the Forum's meetings since that date.

In 2005 the Forum merged with the Christ and the Cosmos Initiative. This had been founded by the Revd Bill Gowland, a past President of the Methodist Conference, with the intention of bringing the latest knowledge of scientific thinking within the orbit of the enquiring layperson.

Thus enlarged, the Forum is open to all who are concerned to relate established scientific knowledge and methodology to religious faith and practice. Complementing its broader objectives, it seeks to:

1) encourage scientists with limited knowledge of religion, and religious people with limited knowledge of science, to appreciate the contributions of both disciplines to human understanding of life in the world

2) provide an interface between academics, active in science-religion work, and public communicators – notably teachers, clerics, and those training future members of these professions.

At every point, the Forum strives to extend recognition that science and religion, properly understood, are not antagonists, but complementary in the quest for truth.

The Forum holds a regular annual conference, plus occasional smaller *ad hoc* meetings, and publishes a twice-yearly journal, *Reviews in Science and Religion*.

At the date of publication the Forum's President is Prof John Hedley Brooke (Lancaster) and its Chairman Prof Neil Spurway (Glasgow).

INTRODUCTION

NEIL SPURWAY

Caution has rarely been the hallmark of the Science and Religion Forum's topic-choices, and it certainly wasn't for the 2007 Conference in the delectable surroundings of Canterbury. The three key words of the conference title, *evolution, mind* and *theology*, combine to form no mean challenge. To start with evolution, it is as prominent in public discussion today as it has ever been. Much of the reason is the recent spread into Britain of resurgent Creationism – a standpoint which no member of the Forum known to me views with anything other than abhorrence. But 2009, the year in which this book appears in print, is of course one in which two major anniversaries of Charles Darwin coincide: the 200th of his birth, and the 150th of the publication of *The Origin of Species*. There will be immense activity in 2009, celebrating and appraising Darwin's extraordinary contribution to humanity's understanding of itself. However the specific matter of the evolution of mind is unlikely to be particularly prominent, and the evolution of the religious mind still less so.

The themes, then, of this 2009 book, drawn from the SRF's 2007 Conference, are theological assessments of evolution generally, and that of mind and the body/mind relationship particularly; but also the evolution of religious thought and – turning the matter round – the implications of an evolutionary standpoint for religious and theological thinking.

Part 1

The evolution of religion

We were extremely pleased to welcome, as our opening speaker, Steven Mithen (Professor of Early Prehistory in the University of Reading), whose 1996 book, *The Prehistory of the Mind*, is a classic presentation of what archaeology and paleo-anthropology can tell us about the origins of science, art and religion in human pre-history. Reassessing, eleven years later, in the light of newer evidence, Steven Mithen was perhaps a little more cautious about which artefacts and grave-contents can be confidently interpreted as religious. Even more certainly, therefore, than a decade ago, we can say that religious responses to the world, insofar as they could leave any trace, have been confined within roughly the last 100,000 years. The second speaker, Celia Deane-Drummond (Director of the Centre for Religion and the Biosciences, University of Chester), responding to Mithen, recalled accounts of the dances of great apes at waterfalls which invite interpretation as essentially religious responses, and the point should no more be dismissed than indications of compassion, or of intuitive physics, among them; however, it is almost certainly not hair-splitting to insist that pre-physics is not yet physics, and pre-religion not yet religion.

Another question might be whether the word "evolution" necessarily applies to the pre-historical development of religion. It should be possible for an old-earth Creationist – one who maintains the separate creation of each species but accepts the time-scale of geological science – to read an account such as Steven Mithen's and feel no need to reject its objective content. But both Deane-Drummond and the contributor of the next chapter, Lluis Oviedo (Professor of Theological Anthropology, Pontifical University "Antonianum", Rome), make clear that the subject, as discussed among 21st C professionals, is steeped in evolutionary thought. Mithen's view of how religious thinking developed derives from evolutionary psychology, though his own addition to this school's modular picture of the mind, that of "cognitive fluidity" enabling interchange between modules, is a more powerful proposal than critics, in this book and elsewhere, assert. Nevertheless, Deane-Drummond presents a telling summary of the scientific arguments against evolutionary psychology itself, not asserting that they are decisive but that they should always be borne in mind. Oviedo, in turn, outlines the wide range of alternative accounts of religious evolution: both different biological ones, arguing for various forms of adaptive advantage, and those which insist that cultural evolution is not just a development of the biological but different in kind.

Broader implications of evolution

The next group of papers considers firstly evolution as such, then its implications for the human capacity to know. Fraser Watts (Reader in Theology and Science at the University of Cambridge), although professionally a psychologist, chose to speak on Darwin's thought and its "gifts to theology" – particularly the gift picked out more than a century before by Aubrey Moore, in Oxford, of emphasizing the ever-presence, the immanence of God, within the day-to-day events which constitute the evolutionary process. As Deane-Drummond had already noted, the idea that divine intervention is to be sought at the points where current science is unable to explain something – the "God of the Gaps" approach – is never met in modern theology, only in the imagined theologies of anti-religious propagandists. Nevertheless, Watts' respondent, Anthony Freeman (Editor of the Journal of Consciousness Studies), insists that a mechanism must be proposed even for the light-touch divine influence on evolution which Watts and Moore uphold, and questions whether either writer is sufficiently clear as to what this might be.

Following this exchange, I myself explore what kinds of theological statement are possible, if one accepts that the human brain/mind is wholly a product of evolution. I conclude that those implying knowledge of events or states of being outside the confines of space and time can have no claim on our acceptance: unlike our concepts of the physical world, or the behaviour of other creatures in it, such propositions have not been tested by our evolutionary survival. This brings down the withering scorn of the respondent, Derek Stanesby (former Canon of St George's Chapel, Windsor). Readers will enjoy his polemic, and will at least grant that I didn't take Editorial advantage and choose a kid-glove antagonist! However I do ask them to consider carefully whether my radically Popperian account of knowledge – constructed wholly in terms of the organism's experiments and the mind's conjectures, surviving or otherwise at the hands of nature – can possibly constitute, as Derek Stanesby asserts, a "return to positivism". Stanesby also denounces evolutionary psychology in general, and its idea of domain-specific mental modules in particular. Now I happen to be sympathetic to this idea, but it is Steven Mithen, not I, who uses it in his paper: at no point in mine does the term "evolutionary psychology" appear.

The relation of mind to body

The next contribution was that of Roger Trigg (currently President of the International Society for the Philosophy of Religion). Against the modern trend, Trigg argues for a dualistic view of the mind-body relation: that they are not, as in the simplistic caricature, two different *stuffs*, but nevertheless two different categories of being. The problem this raises, of course, is what the great neurophysiologist, Sir Charles Sherrington, characterized three quarters of a century ago as "the how of mind's leverage on matter". That is why the majority of modern philosophers and neuroscientists prefer to think not so much of "mind", an entity, as of mental properties or functions, adopting labels like "supervenience" or "dual-aspect monism" to designate their positions.

However, the respondent to this paper, Anne Runehov (Associate Professor of Systematic Theology in the University of Copenhagen) argues for a picture more complex than either dualism or supervenience as usually presented. To place it relative to mainstream 21st C thought, one might characterize her position in terms of two levels of supervenience, or as triple-aspect monism, though these terms are mine, not Anne Runehov's. In this elusive field, one of the best pieces of advice was probably that given me almost 50 years ago by the wonderfully-named philosopher John Wisdom, after he had read my own first, very-young-man's article in the field: "Strive chiefly to be clear what facts about the world you are concerned that your antagonist should not forget."

An approach to Christology

The most ambitious of the invited contributions to the conference is surely that of Jeremy Law (Dean of Chapel at our host institution, Canterbury Christ Church University). Working from an account of human evolution closely paralleling Steven Mithen's, but expressed in his own words and embodying further material, Jeremy Law grasps the nettle of theological anthropology – adding a radically theological dimension to the paleo-anthropological picture, and placing Christ within this framework. The first aspect of any theological perspective has to be that of purpose: here Law's discussion has echoes of that by Watts. Thence recognition of the crucial role of language in the development of *Homo sapiens* leads him to the Word, as represented in the gospel of St John .... and ultimately to the person of Christ, and the "internal conversation" within the Trinity which provides his title.

One of Law's considerations, as he confronts the tension between superficial randomness and hidden purpose, is of the possibility of constrained contingency, imaged by the slight troughs of an uneven snooker table, and this is the aspect of his talk which attracts the main comments of his respondent, Roger Knight, a parish rector in the Diocese of Rochester. Knight points to several inconsistencies in the biblical handling of this idea, and perhaps to one in Law's own use of it. Examples of the first category are the several incidents in which a person – Pharaoh, say, or Judas – is pre-ordained to an evil act and yet held guilty of it. The second is Law's account of Christ's life and death as simultaneously necessary and contingent: can any event be both?

Concluding invited paper

At the end of the conference, the invited main papers were the subject of a masterly overview by Revd Dr Roger Paul (The Church of England's national adviser to the Council for Christian Unity). Among topics which Roger Paul develops much further than the original speakers is the theoretical biologist Stuart Kauffman's investigation of the laws of complexity, whereby increasingly complex systems appear capable of emerging by purely natural processes – Kauffman's account entails no mechanism that is necessarily theological, let alone supernatural. Paul moves on to Steven Mithen's concept of cognitive fluidity, seeing it (rightly in my view) as a much broader capacity than Derek Stanesby believes, and as underlying the whole possibility of metaphorical and symbolic thinking, without which every imaginative expression and every form of abstract thought, from mathematics to theology, would be impossible. Echoing, yet going beyond, my own question, "What can evolved minds know of God?", Roger Paul concludes:

> How can an evolved mind transcend itself? How can it go beyond the structures, ways of thinking and language that the forces of natural selection have shaped? I hope that this book, and the contributions from such a variety of disciplines and convictions, will stimulate further exploration of this question.

Whether or not such transcendence can ever happen, I cannot but share that hope for this book.

Part 2

Contributed papers

Papers offered by people attending the conference are held in parallel sessions, timetabled separately from the sequence of invited papers with which, apart only from Lluis Oviedo's contribution, we have been concerned so far. The selection printed here offers an invaluable set of nuances on the conference theme.

Gavin Hitchcock (University of Zimbabwe) writes of mathematics, the "unreasonable effectiveness" of which (Eugene Wigner's phrase) has puzzled people for generations. Do mathematicians discover eternal truths, "out there" in the world of Platonic ideas, or create the thought-forms in their own, cognitively-fluid minds? Hitchcock seems to incline toward the latter account – as, may I say, do I.

Dick Vane-Wright (late Head of Entomology at the Natural History Museum) considers the rapid growth, in the current generation, of environmental concern. He judges it not only as crucially important for the survival of our species as we know it but also so radical as to represent a significant evolutionary step for the human mind.

Ron Choong (Princeton Theological Seminary, USA) tackles frontally a topic only passingly acknowledged by several of the invited speakers: how the concept of sin stands in the perspective of neuroscience. His focus is on Benjamin Libet's often-cited experiments, indicating that the brain decides on actions before its "owner" knows it has done so.

Finally in this group, Jeremy Swayne, physician and priest, urges a more holistic view of sickness and healing as another necessary step in humanity's view of itself. If achieved, it would surely be an evolutionary change comparable, in both nature and importance, to that discussed by Dick Vane-Wright?

The last chapter is by Revd Dr Sjoerd Bonting (Episcopalian Priest and retired Biochemist from the Netherlands and USA). This *tour-de-force* proved too long for inclusion in the spoken programme, but I am delighted to present it here as a conclusion to our symposium. Different readers will focus on different sections of the paper, which ranges from brain science to the Biblical view of persons. Bonting opens his "Discussion" thus:

> A survey of the biblical data on the mind reveals the position that humans are a body-mind unity and that there is no real distinction between mind and soul. For this reason, I prefer to eliminate the "soul" concept, considering the mind to encompass both our intellectual and spiritual faculties. This view is supported by current neuroscientific insights.

For my own part, I welcome every word of that. Those who do not will find Sjoerd Bonting a tough and terse protagonist.

Body-mind unity

If then, with Editorial arrogance, I were to draw a single moral from these diverse yet admirable – indeed sometimes wonderful – contributions, invited and submitted, it would be that body-mind unity, however expressed, must be the 21st C view of human nature.

So I suggest that even Ron Choong is simplistic, in his assessment of Libet's findings – simplistic in not taking account of the unquantifiable diversity of influences which make up our individual histories and form our characters. In most of our decisions to act we only have time to draw on these unconsciously, but *it is when we do not draw upon them, not when we do*, that we are less than human: unconscious thought is still thought! And the body-mind unit which performs such thought is all the more us, ourselves, because it sets into action before we are aware that it is doing so.

As to the even deeper problems, underlying so many of these papers –
1) the relation of mental functions to the physical brain, and
2) the respect in which Divine will can guide the world without gaps showing in our scientific accounts
– I return again and again to the analogy of my computer. There are so many levels at which its actions can be described. One level is that of electrons, positive holes, and their behaviour at p-n junctions. Next up is that of currents, potentials and resistances, many orders different in one direction than the other. Above that, we soon come to the machine code which instructs these current flows, and the changes of resistance. Above this is the program, written in this instance by an expert in Microsoft Word. But she/he knew absolutely nothing about what I would actually type onto the keyboard ... and an account in those terms shares in turn no elements whatsoever with the thoughts in my mind which lead me to perform my highly-fallible typing. Descriptions at different levels, simultaneously appropriate yet not interchangeable, and not even obviously cross-referring, are the stuff of everyday life. Why is it so hard for us to see that such distinctions between levels of description apply, without gainsay, to the relations between brains and their properties, and between people and their God?

Part 1

CHAPTER ONE

THE PREHISTORY OF THE RELIGIOUS MIND

STEVEN MITHEN

Steven Mithen, FBA, studied at the Slade School of Fine Art and at Sheffield and York Universities (where his topics included computing science) before taking a PhD in Archeology at Cambridge. He then moved to the Department of Archeology at the University of Reading, where he is now Professor of Early Prehistory and Head of the School of Human and Environmental Sciences.

His first book was "Thoughtful Foragers: A Study of Prehistoric Decision Making" (1990). This was followed by probably his most renowned book, "The Prehistory of the Mind: A Search for the Origins of Art, Religion and Science" (1996). Recent publications include "The Singing Neanderthals: The Origin of Music, Language, Mind and Body" (1995).

The essay which follows revisits the position on the rise of religion presented in "The Prehistory of the Mind", testing it against a large body of more recent evidence. It is an expansion of the opening paper of the 2007 conference.

The religious mind

Although any claims about human universals must be made with extreme caution, it is not unreasonable to suggest that types of thought, action and material culture that we classify as "religious" are present in all extant human societies. Individuals within those societies may claim to be atheists, but as far as I know there are no known societies entirely composed of those who have no religious belief. Even atheists may claim to possess feelings about spirituality and engage in activities that an outside observer might categorise as being ritualistic and even religious. The same is likely to apply to all societies documented historically, and those studied archaeologically at least back to the start of the Neolithic, as testified by the pervasiveness of monuments and burials that appear to be of a religious nature. Quite how far back into prehistory religiosity can be traced is a key topic of this essay.

The types of thoughts, actions and material culture in human societies that we describe as religious are enormously diverse. We categorise them together because they share some belief in supernatural power. This often includes the idea that one or more supernatural beings/entities were involved in the creation of the Universe and continue to intervene in the world. By mechanisms such as prayer, meditation, ritual and sacrifice, people believe they can gain a greater understanding of such beings/entities, engage in a form of dialogue with them or seek to influence their interventions in human affairs.

Why should religiosity be so pervasive, perhaps even universal, in human society? There are two possible answers. The first is that there is indeed a supernatural power that was involved in the creation of the universe and may continue to intervene in the world. The diversity of religious thought might simply reflect different but equally valid manifestations of that single truth. Alternatively, this diversity might reflect different degrees of revelation or understanding in which there is progress through time to a more accurate understanding of the nature of the supernatural, which in a broad sense we might describe as the nature of "God". In this regard, "God" may even be choosing to reveal different aspects of itself to people at different stages of their history and in different parts of the world.

The second possible explanation for the pervasiveness of religiosity is that the human mind might simply be prone to believe in the supernatural, even though the origin and on-going activity of the universe and life are explained by entirely natural processes: religion is simply a curious invention of the human mind. As such, it would need to be explained by reference to evolutionary history. This would be the same type of explanation that we use for other universal attributes of human mind such as language and music, all ultimately grounded in natural selection.

How can we differentiate between these two possible explanations? Those who are persuaded of the first tend to invoke the notion of "faith" and contend that it is only by reference to God that there is any meaning to human existence. Those who believe that God and religion are merely artefacts of the human mind point to the progress science is making in explaining aspects of the world that were once attributed to the actions of a supernatural power – most notably the manner in which natural selection has explained the origin of species, including our own species, *Homo sapiens*. But those of a religious persuasion are equally likely to draw on scientific knowledge, pointing to those aspects of the world that science has revealed but seems unable to explain, such as many features of the quantum world.

One source of data that neither party have sufficiently engaged within is that concerning human evolution. The insufficiency does not apply in a general sense, as commentators from Darwin to Dawkins have placed this either implicitly or explicitly at the centre of their opposition to a divine creator, but in terms of the particularities of the fossil and archaeological record. What can these tell us about the origins of the religious mind? In this essay I will review the evidence for human evolution, focussing on the evidence for the earliest forms of religious thought and exploring how this may have become pervasive in human society. First, however, we need to attend to the definition of religiosity.

What is religion?

In general terms, I am simply referring to belief in supernatural agency, whether that is defined as belief in one God, or many Gods, or in spirits, ghosts, animism and so forth. In this essay I am simply using "God" as a short-hand for a religious ideology, with no reference to any particular conception of "God". Pascal Boyer (2001) has usefully re-defined God as an "all-knowing strategic agent", while I also include the attribution of knowledge, will and purpose to inanimate entities as a key element of religious thought. I find the distinction that Harvey Whitehouse (2004) has drawn between "imagistic" and "doctrinal" modes of religiosity useful, especially because these can be broadly related to forms of socio-economic organisation. According to Whitehouse, the imagistic mode consists of the tendency within certain small-scale or regionally fragmented ritual traditions and cults for revelation to be transmitted through sporadic collective action, evoking multi-vocal iconic imagery, encoded in memory as distinct episodes, and producing highly cohesive and particularistic social ties. In contrast, the doctrinal mode of religiosity consists of the tendency within many regional and world religions for revelations to be codified as a body of doctrine, transmitted through structured forms of worship, memorised as part of one's general knowledge and a product of large, anonymous communities.

The recognition of either mode of religiosity from archaeological evidence provides many challenges. The doctrinal mode is more accessible as this tends to create monumental architecture and iconic symbols shared over an extensive area, although identifying these as necessarily of a religious nature may not be as easy as it may initially appear. Moreover, as the doctrinal mode of religiosity is likely to be derivative of a state scale of social organisation of the type that only originated less than 5000 years ago, prehistorians are predominately concerned with identifying religious activity that would fall within Whitehouse's imagistic mode.

Recognising religion in the archaeological record

The most systematic attempt to develop an explicit methodology was by Colin Renfrew in his 1985 book, *The Archaeology of Cult*. This exposed the many steps of inference that an archaeologist must go through when seeking to identify religious activity, often involving the elimination of other explanations for the presence of particular types of artefacts and their particular spatial location and associations within a settlement.

Of the numerous methodological challenges involved in the identification of religious activity from the archaeological record, four can be briefly considered. First, religious thought may have no material representation – it may reside entirely within the mind of an individual. While this cannot be entirely ruled out, I think it is highly unlikely as material objects are frequently, perhaps always, required as cognitive anchors for religious ideas that do not sit comfortably within an evolved mind (Mithen 1998). There remains a dilemma, however, as those material objects might be entirely natural, such as a mountain top or an unmodified stone. As such, although they are visible, an archaeologist is unlikely to appreciate their significance.

Second, religious belief may have material representation but this may be of a nature that does not survive in the archaeological record. All of the objects and structures involved in religious activity might be made from organic materials and hence subject to rapid decay. While this will always be one of the fundamental problems with the reconstruction of past activity and thought, some aspects of it is being alleviated by the developments in archaeological science that continue to enhance the recovery of evidence. The development of isotopic studies of human bone, for instance, has provided archaeologists with information about past diet when no food remains have been preserved, while micro-morphological studies of floor deposits have extracted unprecedented amounts of information about past activities. Both of these can be used to enhance our understanding of past religion, such as by identifying individuals who may have had special diets and areas where non-domestic activities occurred.

A third problem is simply the ambiguity of so much archaeological evidence: objects and structures can easily be misinterpreted as being of religious nature; conversely, items of a religious significance may not be recognised as such – a manger in a stable is likely to be interpreted simply as a feeding place for cattle. A classic example of the former is Neanderthal burial (see Gargett 1989 for a review of the evidence). As I will further discuss below, the discovery that some Neanderthal bodies, of both adults and infants, had been carefully laid within shallow pits inevitably led to proposals about beliefs in an afterlife, while objects found within those so-called graves, such as stone artefacts, animal bones and remnants from flowers, were interpreted as the consequence of graveside ritual. But such burials might be no more than the disposal of "rubbish" in a reasonably hygienic manner and all such artefacts may be part of the rubbish or present for entirely unrelated reasons, such as the parts of flowers coming from the burrowing of rodents – as is likely in the (in)famous case of the Shanidar Neanderthal burials.

A fourth problem to note (there are numerous others, but describing them all would make this essay too depressing) is that our definition of religion might be too restrictive. This derives from the present-day world, or at least that of the recent historical past, which provides us with just a small fraction of the human communities that have lived since the *Homo* genus appeared more than two million years ago. It may be the case that forms of religious belief and action existed in the past that have no modern equivalence; by defining religion on the basis of what we know today, as in the manner of Whitehouse, we risk becoming blind to that of the past.

Human evolution and religious thought

Plio-Pleistocene hominins [1]

As Darwin predicted and as almost a century of fossil discoveries have now demonstrated, human origins are found in Africa. There were numerous species of bipedal primates in Africa between six and two million years ago, these being descendants of the common ancestor we share with the chimpanzee (see Lewin & Foley 2004 for details about hominin fossils relating to this and later sections). These hominins display considerable morphological variation, which most likely relates to the exploitation of specific niches within the African landscape. They have been placed into three genus, *Ardipithecus*, *Australaopithecus* and *Homo*,

---

[1] Hominin is a recent term in paleoanthropology, taking account of DNA-based modifications of the Linnaean classification. However, it includes all species of *Homo* and of *Australopithecus*, so that the lay reader may take it as equivalent, for the purposes of this article, to the Linnaean "hominid". Ed.

with the latter constituted by two purported species, *Homo habilis* and *Homo rudolfensis*. Flaked stone tools are known from at least 2.5 million years ago but in light of the repertoire of tools used by chimpanzees it seems likely that hammer stones, sticks, leaves and other minimally modified materials were used long before flaked stone artefacts appeared.

The two species of *Homo* are characterized by relatively larger brains than the other hominins, up to 650 cc rather than the 450 cc, which is also characteristic of chimpanzees today, together with smaller teeth and a flatter face. These features may be within the range of variation for this grade of hominin without necessarily indicating any evolved cognitive or linguistic abilities. The key problem we face with assessing the significance of brain size is the rarity of the post-cranial skeletons for earliest *Homo*, which leaves open the possibility that the relatively large brains are simply a product of large body size. Indeed, a strong case can be made for reclassifying *H,habilis* and *H.rudolfensis* as australopithecines, and identifying *H.ergaster*, appearing by 1.8 million years ago, as the first member of the *Homo* genus (Wood & Collard 1999).

I cannot find any archaeological evidence within this grade of hominin that would suggest the presence of religious ideas. Their behaviour is readily understood from the perspective of comparative primatology – these hominins are simply relatively large brained and bipedal primates, making a rather more extensive use of stone artefacts than is known among living primates today. While there remains some debate concerning the selective pressures that led to the evolution of bipedalism, more extensive tool use and a meat-based diet, there is no reason to believe that these cannot be explained within the framework of evolutionary ecology as is used to understand the behaviour of animals today (e.g. Aiello & Wheeler 1995).

The evolution of theory of mind

There are, however, three features of these early hominins that may be relevant to the appearance of religious thought at a later stage in human evolution. First, they appear to have been living in larger social groups, this most likely being an adaptive response to the risk of predation in relatively open environments (Aiello & Dunbar 1993). This in turn is likely to have created selective pressures for two cognitive attributes that appear unique to the *Homo* genus and to have been evolutionary related. One of these is what psychologists call "theory of mind" abilities; in simple terms this is being aware that other individuals have beliefs and desires, and that those beliefs and desires might be different to one's own

– although full definitions are problematic (Carruthers & Smith 1996). There is a substantial debate as to whether Theory of Mind capabilities are present in chimpanzees, and if they are how they differ to ours today (e.g. Povinelli 1993, 1999; Tomasello, Call & Hare 2003). The importance of this cognitive ability, or probably this package of abilities, is that it allows one to predict the behaviour of other individuals and consequently facilities both competition (via social manipulation) and co-operation: those individuals who had theory of mind capacities were at a reproductive advantage.

Closely allied to the evolution of theory of mind is that of enhanced communication. Dunbar (1996) has argued that verbal communication would have begun to function as a means of "social grooming": vocalizations would have been used as a relatively cheap means (in terms of time) to develop social relationships, gradually replacing the use of physical grooming for the same end as used among living primates. Enhanced verbal communication would have also been used as a means to manipulate the emotions and hence behaviour of other individuals, this being made possible because of the evolution of theory-of-mind capabilities (Mithen 2000; 2005).

While these first stages in the evolution of theory of mind and linguistic capabilities would have made early hominins a different kind of primate to their own immediate ancestors and to those non-human primates extant today, there are no traces of the distinctive cultural attributes of *Homo sapiens*. There is no evidence for burial, art, architecture and so forth. We are, of course, dealing with archaeological sites that are between 1.5 and 2.5 millions of years old and hence must take into account the problems of preservation and discovery. But there is no hint of anything "cultural" beyond stone artefacts, and certainly nothing that we might wish to place within the category of religious.

Early Humans: from *Homo ergaster* to *Homo neanderthalensis*

H.ergaster, appearing at around 1.8 million years ago, may mark an evolutionary transition to a type of hominin for which behavioural analogies with living primates are of limited value. With a near-fully modern stature and bipedal gait, this species is most likely the first to disperse out of Africa. Brain size reached up to 900 cc, although some specimens show the maintenance of relatively small brain capacities – those from the site of Dmanisi in Georgia were no more than 650cc (Gabunia *et al.* 2000; Lordkipanidze *et al.* 2000). The Asian lineage of this species evolved into *Homo erectus* and appears to have made at least one water crossing on rafts to reach Flores Island by 850,000 years ago (Morwood *et al.* 1998).

In Europe, the middle Pleistocene is marked by a succession of hominins that are claimed to constitute at least three species: *Homo antecessor*, *H.heidelbergensis* and *H.neanderthalenis* (Lewin & Foley 2004). These may form a single evolving lineage or multiple dispersals into Europe of species that evolved from *H.eragster* in Africa.

These hominins appear to show a gradual increase of brain size until a capacity equivalent to, and in some cases exceeding that of *Homo sapiens* is attained in late Pleistocene specimens. Stone artefacts appear to increase in technological complexity from Oldowan-like flakes associated with *Homo antecessor*, Acheulian handaxes with *H.heidelbegensis* and levallois technology with *H.neanderthalensis*. Examples of bone and wooden tools are exceedingly rare, but the discovery of the Schöningen spears (Thieme 1997) indicates that this is most likely a consequence of the rarity of both preservation and discovery. Traces of structures and non-utilitarian artefacts are effectively absent, with the few ambiguous examples, claimed by some, merely emphasizing their extreme rarity and unsophisticated nature.

Can we find anything in the archaeological record of these "Early Humans" that suggest religious thought? Some claim that we can, and point to two lines of (exceedingly uncommon) evidence: objects that appear not to have any utilitarian function, and burials.

Contentious symbolic artefacts

There are no unambiguous examples of visual symbols related to this grade of hominin; even if there were, the connection with religious thought would still need to be established. The most provocative artefact is a small piece of stone (c. 3cm) from the site of Berekhat Ram in Israel that dates to c. 250,000 years ago and is claimed by some to be a figurine, an image of a female (D'Errico & Nowell 2000). Superficially, there is a resemblance: one can visualize a head, arms and bust; moreover microscopic analysis has demonstrated that the stone has been deliberately incised with a stone blade. But whether the female form is simply in the "eye of the beholder", as I believe it is, or has been deliberately imposed, remains contentious. Those who believe that the Berekhat Ram "figurine" is the earliest known representation of the human form have drawn analogies with the significantly larger Venus Figurines of the Upper Palaeolithic, these having been made between 30,000 and 20,000 years ago by modern humans in ice age Europe. As I will elaborate on below, those Venus Figurines unquestionably testify to religious thought. But there is absolutely no basis for any link with the tiny piece of stone from Berekhat Ram.

Other than this object, the closest we appear to get to any form of art object made by Early Humans are incised pieces of bone from the site of Bilzingsleben, dating to around 350,000 years ago (Mania & Mania 1988). These have been claimed to be of symbolic significance because the incisions appear to be ordered; indeed on one piece of bone they form a neat set of parallel lines. There are potential utilitarian explanations, such as that the bones had been used as "cutting boards" for meat and plant, resulting in the types of incisions that we find on our bread-boards at home. Whether this is adequate remains unclear, and to me the lines look more ordered and deliberate than would be expected from accidental marking. Two enormous claims are required to suggest that such incised artefacts suggest a religious mind. First, one must claim that the marks have a symbolic significance, and second, one must claim that they are symbolising some form of religious entity. Neither claim can be substantiated on the basis of present evidence. Nevertheless, extravagant assertions have been made by certain archaeologists and commentators about the Bilizingsleben hominins, suggesting that they had pagan-like religious beliefs. There is no evidence for this.

The evidence from burials

Burials might provide a more promising source of evidence for those who wish to find religiosity in early prehistory. The earliest known burials are those in the caves of Skhul and Qafzeh at Mount Carmel, Israel that date to between 100,000 and 80,000 years ago (Vandermeersch 1970). These are of a lineage of *Homo sapiens* that most likely dispersed out of Africa during the last interglacial (c. 125,000 years ago). The burials have grave goods in the form of animal parts, such as the jaw of a wild boar and antlers of deer that have been carefully positioned with the bodies. Red ochre had also been used during what ought to be described as a burial ritual (Hovers *et al.* 2003).

These burials are of *Homo sapiens* – although a lineage that appears to have gone extinct, making no contribution to the modern gene pool. Of greater interest to the concerns of this essay are the burials of Neanderthals (*H neanderthalensis*) that are found between 65,000 and 30,000 years ago in the Near East and Europe. There are numerous examples of such burials, the most famous being those of the Shanidar cave in Iraq, Amud cave in Israel and La Ferrassie in France (see Gargett 1989, 1999 for extensive discussion of these and other Middle Palaeolithic burials). There has been a long and detailed debate within archaeology about such burials. Three questions are most pertinent. First, are they deliberate burials or simply accidental, such as from rock falls within caves? While some might be of the latter type, there are numerous examples where it is clear that a pit has been excavated and a body carefully lain within. Second, do these burials simply represent the hygienic deposition of a rotting corpse? While this is an intriguing idea, two factors mitigate against it. One is that the Neanderthals are not known to be particularly tidy. Their living sites appear to be replete with the rubbish from the tools they have made and animals they have butchered; without any other evidence for systemic rubbish disposal, it seems far fetched that this would have been applied to human bodies. The other reason is that there are much easier and safer ways to dispose of a dead body than by burying it in the ground, especially within a cave that one is continuing to use. Dead bodies can be slung into rivers or simply left for the vultures, hyenas and eventually beetles to take away.

The third question is whether any grave goods are present within the Neanderthal burials? Numerous claims have previously been made, notably that of wreaths of flowers laid across the bodies in Shanidar Cave. But none of these claims have stood up to scrutiny. Any artefacts or animal bones found within the graves are more reasonably interpreted as incidental contents of the material used to back-fill the grave. The claim for flowers at Shanidar was made on the basis of high frequencies of flower pollen within the grave fill. This is now thought to be a consequence of burrowing rodents that had used flower parts to line their nests.

Without grave goods, or indeed any signs of graveside ritual, it is difficult to make a claim that Neanderthals burials are related to religious ideas, such as belief in an after-life. A more reasonable interpretation is simply that they had buried members of their community, ranging from very young infants to mature adults, simply because they had loved them during life and wish to care for their bodies after death (Mithen 2005). We know that the Neanderthals did care for each other as there are examples of individuals with healed injuries that would have required support during their healing process. Indeed, there is no reason to think that the Neanderthals were any less empathetic and emotional than we are today; they may have been more so because their populations were constantly teetering on the edge of extinction and hence the loss of any individual may have threatened the on-going existence of the whole community. In this regard their theory-of-mind capabilities, that had begun to evolve within Plio-Pleistocene hominins were most likely as advanced as our own.

Explaining the absence of religious thought before *H.sapiens*

In light of the absence of any convincing artworks, symbols, grave goods or ritual behaviour of any kind for not only the Neanderthals but also for *Homo ergaster, erectus* and *heidelbergensis*, there is no basis for arguing that these hominins had any form of religious belief. We must, of course, always be cautious in light of issues about preservation in the archaeological record and the fact that religious ideas could conceivably have had no material representation (although I think the latter is highly unlikely).

This absence of evident religiosity is intriguing as in so many other ways these early Humans, and especially the Neanderthals, are so similar to modern humans for whom religion appears pervasive, if not universal. The Neanderthals had brains as large as ours today, made complex stone artefacts displaying high levels of technical skill, and could adapt to a wide range of environmental conditions (see Stringer & Gamble 1993 and Mellars 1996 for overviews of the Neanderthals). Their vocal tracts were very similar to modern humans, indicating that there was no anatomical reason why they could not have possessed language. Some form of advanced communication must have been essential in light of what we know about their lifestyles (Mithen 2005).

There are, however, some striking differences between Neanderthals and modern humans in addition to the apparent absence of a religious mind. There is, for instance, an astonishing absence of cultural innovation: the Neanderthals appear to have made the same types of stone artefacts year after year for millennia, even through periods of dramatic environmental change. They did respond to changes in raw material supply and the resources they had to exploit, but only in a highly limited manner. For a population that appears to have been under significant levels of adaptive stress, it is surprising that there appears to have been no invention of projectile technology.

My own explanation for the striking similarities and differences between Neanderthals and *Homo sapiens*, and one that would explain an absence of religious thought, is that their minds had different forms of architecture (Mithen 1996). The Neanderthals appear to have had a "domain-specific" mentality. By this I mean that they may have had stores of knowledge and ways of thinking about the social world, about tools, and about the natural world, as sophisticated as that of modern humans, but these were largely isolated from each other. So although their theory-of-mind capability is likely to have been as advanced as that of modern humans, the Neanderthals could not apply these to animals as we do habitually when we engage in anthropomorphic thinking. For the Neanderthals the human, animal and material worlds were thought about in isolated cognitive domains which imposed severe constraints on their creative abilities. As I will explain below, it was from the merging of those domains that religious ideas emerged in *Homo sapiens*.

With regard to communication, the cultural stasis and absence of visual symbols suggests that the Neanderthals lacked language in the sense of words and grammar. Words, after all, are merely symbols and had the Neanderthals been able to manipulate audible symbols I find it inconceivable that they would not have also been making and manipulating visual symbols. Among modern humans language is a motor for cultural change: by talking about our tools, describing, comparing and contrasting them, we learn how to improve them. If Neanderthals had language, how could they have stayed making the same basic types of tools for so many thousands of years? Nevertheless, the requirements of competition and cooperation in Neanderthal society would have required some form of communication far more advanced than we find in non-human primates today. I have suggested that they had a holistic form of communication – a relatively fixed set of messages rather than words – that made extensive use of variations in pitch, rhythm and tone. This was, I have argued, common to Early Humans and within the African lineage provided the root for not only language but also music (Mithen 2005).

The origin of modern humans

Within Africa there is evolutionary continuity from *Homo ergaster* to *sapiens* (McBrearty & Brooks 2000), with the earliest specimens of the latter dating to c. 200,000 years ago (McDougall, Brown & Fleagle 2005), a date that effectively coincides with an estimate for this species' appearances from the study of modern-day genetic diversity (Ingman *et al.* 2000; Jobling,, Hurles & Tyler-Smith 2004). While *H.sapiens* specimens from Mount Carmel indicate initial dispersal out of Africa prior to 100,000 years ago (Lahr & Foley 1994), the genetic evidence indicates that it was after 60,000 years ago that major dispersals into Asia and Europe occurred, and that it was these that gave rise to the extant populations today (Ingman *et al.* 2000).

Immediately prior to such dispersals we find evidence in South Africa for new types of material culture, often assumed to reflect the appearance of symbolic thought and language. Most notable are the incised ochre nodules and shell beads from Blombos Cave dating to 70,000 years ago (Henshilwood *et al.* 2002, 2004), while red ochre is prevalent in Middle Stone Age deposits at South African sites reaching back to 100,000 years ago (Knight, Powers & Watts 1995). The most recent discovery, that from Pinnacle Point Cave, indicates that ochre may have been deliberately used as a colouring agent as far back as 165,000 years ago (Marean *et al.* 2007).

The dispersals of *H.sapiens* out of Africa resulted in the colonisation of Australia by at least 30,000 years ago, and possibly by 59,000 years ago, and that of Europe by 40,000 years ago. The latter is associated with major technological innovations characterised by the Aurignacian culture that appears to arise in the Near East. The Neanderthals may have attempted to imitate the culture of the incoming *H.sapiens* (D'Errico *et al.* 1998) but they were either out-competed for resources or unable to survive the major climatic fluctuations of the late Pleistocene (D'Errico & Sanchez Goni 2003). By 30,000 years ago, the *H.sapiens* in Europe were engaging in cave painting, carving intricate bone figurines and making elaborately decorated burials. Similarly, the *H.sapiens* in Australia and Africa were likely to be engaged in both abstract and figurative rock art at this date.

The Neanderthals became extinct soon after 30,000 years ago. Precisely when *H.erectus* in Asia became extinct remains unclear; a similar date is likely, although *H.floresiensis* survived on Flores island until a mere 13,000 years ago (Morwood *et al.* 2004). The climatic deterioration of the last glacial maximum some 20,000 years ago caused *H.sapiens* to abandon northern landscapes that became polar desert, and those areas in low latitudes that became extremely arid. As global warming began, *H.sapiens* re-colonised those landscapes and proceeded to colonise the rest of the world either in the late Pleistocene[2] or early Holocene - the far north, the Americas and the islands of the Pacific (Mithen 2003).

It was only in the Holocene, beginning a mere 10,000 years ago, that agricultural economies developed, initially in the Near East at c. 9000 years ago and then quite independently at several locations elsewhere in the world – including rice farming in China 7000 years ago, maize and squash in Central America 6000 years ago and domesticated camelids (llamas and alpacas) in the Peruvian Andes by 5000 years ago (Smith 1995; Mithen 2003). Farming provided the economic foundation for the development of towns and within a few thousand years the first "civilisations", within which writing was independently invented.

---

[2] Pleistocene = period of the ice ages, "late" from say 15,000 years ago. Ed.

The first evidence for religious thought

When in this span of time since the origin of *Homo sapiens* soon after 200,000 years ago, can we see the first indications of a "religious mind"? The Mount Carmel burials of Skhul and Qafzeh at 100,000-80,000 provide strong evidence, I believe, that there is now thought about "another world". These burials are strikingly different to those of the Neanderthals and suggest significant graveside ritual.

Whether or not the incised pieces of red ochre from Blombos Cave and the seemingly widespread use of that pigment in the Middle Stone Age of South Africa indicate the presence of a religious mind is unclear. Red ochre is certainly used extensively in the religious rituals of the historic San of South Africa. Moreover, we know from ethnographic studies of Aboriginal art that even simple geometric designs as found on the Blombos artefact can have highly complicated meanings. I strongly suspect that those early modern humans of South Africa were engaging in religious thought. But it is only with the first representational art of the Upper Palaeolithic in Europe, dating to c. 35,000 years ago, that we can be entirely confident that human minds are having ideas about supernatural beings.

Within this art, lasting until the end of the last ice age and predominantly found in South West Europe, there are striking images of entities that appear to be part human and part animal (see Bahn & Vertut 1997, and White 2003 for overviews of Upper Palaeolithic art). The most famous examples are the lion-man from Hohenstein Stadel, a carving from mammoth ivory of a man's body with a lion head, and the bison/woman figure from Chavet Cave – a painting of a female torso with bison shoulders and head. Although these may simply represent people wearing animal skins, it would be perverse to suggest that they did not represent beings of the ice-age spirit world. Indeed, the whole of Upper Palaeolithic art is most likely closely related to a mythological world to which we have no access today other than via the paintings and sculptures. When making this claim we should take the context of the paintings into consideration, some of them being found deep below ground.

The interpretation of the religious thought and action behind Upper Palaeolithic art by expert scholars such as Jean Clottes and David Lewis-Williams (Clottes. Lewis-Williams & Hawkes 1998; Lewis-Williams 2002) are consistent with Whitehouse's concept of imagistic religious. These include traumatic or violent initiation rituals, experiences of collective possession and altered states of consciousness. The specific role that the art played in such ice-age religion remains unclear. Similarly, it remains questionable whether we can draw any specific inferences about the nature of ice-age supernatural beings from that art and associated archaeological data.

My view is that the materialisation of the religious ideology of the Upper Palaeolithic in the paintings and sculptures played an active role in the formation of that ideology and its transmission between generations (Mithen 1998). As we are dealing with an imagistic religious mode, that ideology would have been ill-defined: while sharing some basic ideas, each person would have had their own particular conception of the supernatural beings represented in the art. The paintings and carvings should be considered as shared parts of their Upper Palaeolithic minds rather than as mere products or expressions of those minds. This is, of course, no more than Leach (1976, p 37) argued thirty years ago when he explained that we convert religious ideas into material form to give them relative permanence, so that they can be subjected to operations which are beyond the capacity of the brain alone.

The origin of religious thought

While the evidence from Qazeh, Skhul and Blombos Cave remains difficult to interpret, that of the Upper Palaeolithic demonstrates that modern humans of the ice age were able to engage in religious thought. The rock art of a comparable age from Australian and Tasmania can be interpreted in a similar manner. It appears most reasonable to argue that religious thought in fact originated with the appearance of modern humans in Africa c. 200,000 years ago. One might argue here that this was a moment of "divine intervention" – that this was the first moment that "God" chose to begin his revelation to humanity. I don't, however, think such arguments are necessary, but rather that the origin of religious thought at this time can be explained by evolution. More specifically, it derives from the impact that the evolution of fully modern language had on the architecture of the mind.

I have already described how Early Humans possessed a domain-specific mental architecture: their minds were composed of isolated cognitive domains. Modern human minds are different because they have the quality of "cognitive fluidity" – the sharing of knowledge and ways of thinking between different domains. This provides the capacity for metaphor and analogy – the basis for both art and science. I have previously suggested that language provides the vehicle for the flow of knowledge and ways of thinking from one cognitive domain to another (Mithen 1996). That fully modern language, making use of an extensive lexicon and suite of grammatical rules, first originated with modern humans is a widely held view among archaeologists and linguists. There are numerous different lines of evidence to suggest that language did indeed originate at this time, and numerous different explanations for quite how and why it did so. My own view is that language evolved from a complex form of communication that was as much music-like as language-like, as I have argued at length in my 2005 book *The Singing Neanderthals*. However, our concern here is with the impact of language on the nature of thought.

That this was profound has been considered in depth by the philosopher Peter Carruthers (2002, 2006) when considering the "Cognitive functions of language". He argued that the "imagined sentences" we create in our minds allow the outputs from one domain (or module) to be combined with those from one or more others, and thereby create new types of conscious thoughts. The consequence is that we can imagine entities that do not – and cannot – exist in the real world. Perhaps the most pervasive consequence of cognitive fluidity has been for modern humans to imagine that all events have intentionality and meaning. A key element of the pre-modern human social intelligence was to infer the intentions behind another person's actions – the theory-of-mind ability. Once that way of thinking became accessible to natural history and technical intelligence, people began to wonder what the intention was behind natural phenomenon, such as thunderstorms, earthquakes and the sight of rare animals. As Guthrie (1993) has argued, such anthropomorphising of the natural and inanimate world is a pervasive feature of religious thought, and perhaps a defining characteristic. A more specific consequence of cognitive fluidity was the ability to imagine the existence of supernatural beings.

The cognitive anthropologist Pascal Boyer (1994, 2001; Boyer & Ramble 2001) has made extensive cross-cultural studies of such beings. Along with Mithen (1996) he argues that they combine features of different cognitive domains (or in his terminology, "intuitive ontologies"). Ghosts, for instance, are often envisaged as being just like humans, except that they can pass through solid objects in the manner that sound or vibrations are able to do. A statue of the Virgin Mary is an inanimate object but has somehow acquired the psychological propensities of a person because it/she can hear prayers. Through his anthropological and experimental studies, Boyer has found that the types of supernatural beings that are most prone to be believed and most resilient to cultural transmission are those that are only "minimally counterintuitive". They must have counter-intuitive properties to have salience - such as a "man" who can rise from the dead, does not need to feed, or lives in some other reality such as "heaven" or the Dreamtime. On the other hand, they must have sufficient contact with an evolved cognitive domain for them to be conceived at all and to be transmitted across generations: thus the Classical Greek Gods, Aboriginal Ancestral Beings, and the vast majority of supernatural beings recorded by anthropologists, have often behaved and thought like "real" people.

The chain of argument is, therefore, that the origin of language led to cognitive fluidity and this enabled the mental conception of supernatural beings (along with many other new types of ideas) and hence formed the basis for religious thought. On the basis of this argument, all modern humans have a propensity for religious thought. This is simply a spin-off from the evolution of language and the resulting cognitive fluidity of the modern mind. Once present, the propensity for religious thought could be used and exploited in many different ways. That would take us onto the history of religion itself – which is beyond the scope of this essay.

Summary

I began this essay with the claim that a propensity to religious thought is universal among *Homo sapiens*. This is in some regards the "natural state" of humankind, as it appears to involve a greater cognitive effort to be an atheist than it does to be a believer. I suggested that there were two possible explanations. Either human minds are in various states of knowledge about the true existence of a "God" that/who is in the process of revelation, possibly in various different manifestations. Alternatively, the propensity for religious thought is simply a by-product of an evolved mind, no more than a curious feature of human mentality – although one that has had profound consequences for human history.

One way to attempt to differentiate between these possible explanations is to interrogate the archaeological record as to when and why religious thought first appeared during human evolution. This is a demanding task as it involves numerous methodological challenges about how religious thought and action can be recognized archaeologically. A great deal of the evidence is highly ambiguous to interpret, while a great deal more might be simply missing owing to a lack of preservation.

With these caveats in mind, I have provided an overview of the archaeological evidence for religious thought, often combining this with my own interpretations. I conclude that religious thought is uniquely associated with *Homo sapiens* and arose as a consequence of cognitive fluidity, which was in turn a consequence of the origin of language. In this regard, there appears to be no need to invoke a moment of divine intervention that initiated the start of a revelation. For me, therefore, there is no supernatural, no God to be revealed.

Acknowledgement

I am grateful to Neil Spurway for inviting me to talk at the 2007 Science and Religion Forum meeting at Canterbury Christ Church University and to contribute to this volume.

References

Aiello, L.C. & Dunbar, R.I.M. 1993. Neocortex size, group size, and the evolution of language. *Current Anthropology* 34, 184-193.

Aiello, L.C. & Wheeler, P. 1995. The expensive-tissue hypothesis. *Current Anthropology* 36, 199-220.

Bahn, P.G. & Vertut, J. 1997. *Journey through the Ice Age.* London: Weidenfeld & Nicolson.

Boyer, P. 1994. *The Naturalness of Religious Ideas: A Cognitive Theory of Religion.* Berkeley: University of California Press.

—. 2001. *Religion Explained: The Evolutionary Origin of Religious Thought.* New York: Basic Books.

Boyer, P. & Ramble, C. 2001. Cognitive templates for religious concepts: cross cultural evidence for recall of counter-intuitive representations. *Cognitive Science* 25, 535-564.

Carruthers, P. 2002. The cognitive functions of language. *Brain and Behavioral Sciences* 25, 657-726.

—. 2006. *The Architecture of the Mind.* Oxford: Clarendon Press.

Carruthers, P. & Smith, P. (eds) 1996. *Theories of Theories of Minds.* Cambridge: Cambridge University Press.

Clottes, J., Lewis-Williams, D. & Hawkes, S. 1998. *The Shamans of Prehistory: France and Magic in the Painted Caves.* New York: H. Abrams Inc.

D'Errico, F. & Nowell, A. 2000. A new look at the Berekhat Ram figurine: Implications for the origins of symbolism. *Cambridge Archaeological Journal 10*, 123-167.

D'Errico, F. & Sanchez Goni, M.F. 2003. Neanderthal extinction and the millennial scale climatic variability of OIS 3. *Quaternary Science Reviews* 22, 769-788.

D'Errico, F., Zilhão, J., Julian, M., Baffier, D. & Pelegrin, J. 1998. Neanderthal acculturation in Western Europe. *Current Anthropology* 39, S1-S44.

Dunbar, R.I.M. 1996. *Gossip, Grooming and the Evolution of Language*. London: Faber & Faber.
Gabunia, L., Vekua, A., Lordkipanidze, D. *et al.* 2000. Earliest Pleistocene hominid cranial remains from Dmanisi, Republic of Georgia: taxonomy, geological setting and age. *Science* 288, 1019-1025.
Gargett, R.H. 1989. Grave shortcomings: the evidence for Neanderthal burial. *Current Anthropology* 30, 157-190.
—. 1999. Middle Palaeolithic burial is not a dead issue: the view from Qafzeh, Saint Césaire, Kebara, Amud, and Dederlyeh. *Journal of Human Evolution* 37, 27-90.
Guthrie, S. 1993. *Faces in the Clouds: A New Theory of Religion*. Oxford: Oxford University Press.
Henshilwood, C.S., d'Errico, F., Yates, R., *et al.* 2002. Emergence of modern human behavior: Middle Stone Age engravings from South Africa. *Science* 295, 1278-1280.
Henshilwood, C.S., d'Errico, F., Vanhaeren, M., van Niekerk, K., Jacobs, Z. 2004. Middle stone age shell beads from South Africa. *Science* 304, 404.
Hovers, E., Ilani, S., Bar-Yosef, O. & Vandermeersch, B. 2003. An early case of color symbolism: ochre used by early modern humans in Qafzeh Cave. *Current Anthropology* 44, 491-522.
Ingman, M., Kaessmann, H., Paabo, S. & Gyllensten, U. 2000. Mitochondrial genome variation and the origin of modern humans. *Nature 408, 708-13.*
Jobling, M.A., Hurles,M.E. & Tyler-Smith, C. 2004. *Human Evolutionary Genetics*. New York: Garland Publishing.
Knight, C., Powers, C. & Watts, I. 1995. The human symbolic revolution: a Darwinian account. *Cambridge Archaeological Journal* 5, 75-114.
Lahr, M.M. & Foley, R. 1994. Multiple dispersals and modern human origins. *Evolutionary Anthropology* 3, 48-60.
Leach, E. 1976. *Culture and Communication*. Cambridge: Cambridge University Press.
Lewin, R. & Foley, R.A. 2nd ed. 2004. *Principles of Human Evolution*. Oxford: Blackwell Publishing.
Lewis-Williams, D. 2002. *The Mind in the Cave*. London: Thames & Hudson.
Lordkipanidze, D., Bar-Yosef, O. & Otte, M. 2000. *Early Humans at the Gates of Europe*. Liège: ERAUL 92.
Mania, D. & Mania, U. 1988. Deliberate engravings on bone artefacts of *Homo erectus*. *Rock Art Research* 5, 91-107.

Marean, C.W., Bar-Matthews, M., Bernatshez, J *et al.* 2007. Early human use of marine resources and pigment in South Africa during the Middle Pleistocene. *Nature* 449, 905-908.

McBreaty, S. & Brooks, A. 2000. The revolution that wasn't: a new interpretation of the origin of modern human behavior. *Journal of Human Evolution* 38, 453-Mellars, P. 1996. *The Neanderthal Legacy.* Princeton: Princeton University Press.

McDougall, I., Brown, F.H. & Fleagle, J.G. 2005. Stratigraphic placement and age of modern humans from Kibish, Ethiopia. *Nature* 433, 733-6.

Mellars, P. 1996. *The Neanderthal Legacy. An Archaeological Perspective from Western Europe.* Princeton, N.J.: Princeton University Press.

Mithen, S.J. 1996. *The Prehistory of the Mind: A Search for the Origins of Art, Science and Religion.* London: Thames & Hudson.

—. 1998. The supernatural beings of prehistory and the external storage of religious ideas. In Cognition and Material Culture: *The Archaeology of Symbolic Storage* (eds. C. Renfrew & C. Scarre), pp. 97-106. Cambridge: McDonald Institute of Archaeological Research.

—. 2000. Palaeoanthropological perspectives on the theory of mind. In *Understanding Other Minds* (ed. S Baron-Cohen, H. Tager-Flusberg & D.J. Cohen) pp. 488-502. Oxford: Oxford University Press.

—. 2003. *After the Ice: A Global Human History, 20,000-15,000 BC.* London: Weidenfeld & Nicolson.

—. 2005. *The Singing Neanderthals: The Origin of Music, Language, Mind and Body.* London: Weidenfeld & Nicolson.

Morwood, M.J., O'Sullivan, P.B., Aziz, F. & Raza, A. 1998. Fission-track ages of stone tools and fossils on the east Indonesian island of Flores. *Nature* 392 173-176.

Morwood, M.J., Soejono, R.P., Roberts, R.G., Sutikana, T.,Turney, C.S.M., Turney, Westaway, K.E., Rink, W.J., Zhao, J.-x., van den Bergh, G.D., Rokus Awe Due, Hobbs, D.R., Moore, M.W., Bird, M.I. & Fifield, L.K. 2004. Archaeology and the age of a new hominin from Flores in eastern Indonesia. *Nature* 431, 1087-1091.

Povinelli, D.J. 1993. Reconstructing the evolution of the mind. *American Psychologist* 48, 493-509.

—. 1999. *Folk Physics for Apes.* Oxford: Oxford University Press.

Renfrew, A.C. (ed.) 1985. *The Archaeology of Cult, the Sanctuary at Phylakopi*, London: British School at Athens and Thames & Hudson.

Smith, B.D. 1995. *The Emergence of Agriculture.* New York: Scientific American Library.

Stringer, C. & Gamble, C. 1993. *In Search of the Neanderthals.* New York: Thames & Hudson.

Thieme, H. 1997. Lower Palaeolithic hunting spears from Germany. *Nature* 385, 807-810.
Tomasello. M., Call, J. & Hare, B. 2003. Chimpanzees understand psychological states - the question is which ones and to what extent. *Trends in Cognitive Sciences* 7, 153-156.
Vandermeersch, B. 1970. Une sépulture moustérienne avec offrandes découverte dans la grotte de Qafzeh. *Comptes Rendus Hebdomadaires des Séances de l'Académie des Sciences* 270, 298-301.
White, R. 2003. *Prehistoric Art: the Symbolic Journey of Humankind.* New York: Harry N. Abrams, Inc.
Whitehouse, H. 2004. *Modes of Religiosity: A Cognitive Theory of Religious Transmission.* Walnut Creek, CA: Altamira Press.
Wood, B. & Collard, M. 1999. The human genus. *Science* 284, 65-71.

CHAPTER TWO

WHENCE COMES RELIGION?
MITHEN ON PREHISTORY AND MIND

CELIA DEANE-DRUMMOND

Professor Celia Deane-Drummond has worked for many years at the interface of biological science and theology, after gaining first degrees (Cambridge, Manchester) and doctorates (Reading, Manchester) in both plant science and theology. She writes prolifically, her most recent books being "Genetics and Christian Ethics" (2006); "Future Perfect: God, Medicine and Human Identity", edited with Peter Scott (2006) and "Teilhard de Chardin on People and Planet" (2006). She holds a chair in Theology and the Biological Sciences at the University of Chester and is Director of the Centre for Religion and the Biosciences there, at the same time as being the mother of two young children.

In this detailed response to Prof Mithen's paper, she challenges the evolutionary psychology on which he bases so much of his thought, both for the models of mind which it employs and for its assumption that every habit of the modern human mind must have had adaptive advantage at an earlier stage in evolution. At a number of points, also, she notes that recent, science-aware theology is subtler than Mithen assumes.

Initial reactions

Anyone encountering the work of Stephen Mithen for the first time cannot help but find it a fascinating and gripping account of the earliest origins of our hominid ancestors. It is, of course, the extension of this analysis to account for the earliest religious instincts that one can anticipate being the most controversial for theologians, though I would like to suggest it is also a way of considering our past in evolutionary terms that provides one of the deepest challenges to ways of thinking about our human identity, including the origins of morality and religion. I found his paper particularly illuminating in as much as it brought to the surface in a vivid way the particular problems encountered in this kind of research, in respect of the traces or lack of traces of religious belief that are left behind. He is also honest enough to suggest when there is no evidence for religious activity, or where the evidence is thin. In this he provides a valuable check on those archaeologists who have been tempted to speculate about the religious beliefs of early hominins from very scant evidence. The lack of apparent religious belief even among Neanderthals, who in so many other respects were like modern humans, speaks of a lack of the imaginative capacity that is perhaps necessary for religious sensitivity.

Theologians have much to learn from this painstaking and careful approach to the study of our distant ancestors. Unlike many theologians who often presuppose that moral rules are coded into a framework of particular religious beliefs, evolutionary biologists more often than not describe basic moral rules as also existing in non-humans, with particular religious instincts, such as for symbol making, being characteristic of *Homo* species. I am in support of Mithen's most fundamental concern, namely, to suggest that religious belief is, at least, common in all human societies, and, if the archaeological record is correct, is pervasive as far back as Neolithic human societies. I also have no problem with a broad anthropological analysis of religion in as much as he follows authors like Harvey Whitehouse in trying to discern distinctive patterns in religion that are either imagistic or doctrinal. He also suggests that the more imagistic mode is more likely to be characteristic of religious pre-history. The question that he asks, namely, "Why is religion so pervasive in human societies?", is also a fascinating question, whatever religious or non-religious belief one happens to hold dear.

Two over-simplifications

In engaging in more detail with Mithen's work, I would like to pick up some underlying areas of discussion that I believe are crucial to the development of his account. The first is that he draws heavily on theories of the mind developed in evolutionary psychology. In common with many writers influenced by this line of inquiry, at the start of his paper he names two possible scenarios: either there is a God who is creator, intervening in nature, and religious diversity reflects different manifestations or facets of a single truth, or there is simply a general tendency for the mind to be religious, explained by entirely natural processes, grounded in natural selection. The contrast set up, of course, invites any reader who is not already committed to religious belief to opt for the second alternative. Yet there is no need to posit God as one who intervenes supernaturally in the way Mithen suggests. Theologians, at least contemporary ones who are familiar with evolutionary biology, do not envisage God as a divine intervener in nature, but as a God who works *through* the natural forces of evolution.[1] There are also different degrees of identification with creation, from pantheism, through panentheism to deism. It is also poor theology to point to gaps in our knowledge and seek to find God where there is no science. The possibility that religion might have a natural basis in evolutionary terms is also not threatening to contemporary theology, for it is possible to hold to such religious beliefs and also believe in evolutionary psychology – a view held by Justin Barrett (2004).

The question that is really to be addressed is whether "religion is simply an invention of the human mind", for just because in a very loose way religious capacity is *enabled* by the kinds of brains that have evolved, does not necessarily mean that religion can be *explained* by evolutionary capability, any more than having neural networks *explains* mental function. There is a difference, then, between a causal explanation, and enablement – a point I will pick up again below. Evolutionary psychology conflates the two. Mithen in fact goes further than the basic "Swiss army knife" proposal, of the earlier pioneers Leda Cosmides and John Tooby, in as much as he contemplates considerable flexibility of mental response. However, rather than go down the route of total cognitive fluidity, instead he understands the mind as existing much as a cathedral, with passages between the different mental modules, allowing communication between them. Of course, for a theologian, the analogy with a cathedral is instantly appealing, and certainly the idea allows for a story to be told about how early and late humans differed in their abilities to think in complex ways, seemingly accounting for some empirical observations about the differences in cultural behaviour, including, perhaps, religious behaviour.

---

[1] The idea of a non-interventionist God has been characteristic of the science and religion discourse for a number of years, and has been perhaps most in evidence in a series of books published jointly by Vatican City and the Centre for Theology and the Natural Sciences. The most relevant volume in this context is Russell, Stoeger & Ayala (1998); see especially Peacocke's paper there.

Scientific objections to Evolutionary Psychology

I think it is important to be aware of the objections to evolutionary psychology that have surfaced in the *scientific* literature, even if, to a large extent, evolutionary psychology has carried on in its path, largely ignoring any resistance, and to some extent proving itself by establishment of a wide range of undergraduate textbooks and resources for students. One might think, looking at the extent of the literature available, that this is truly a science come of age (Buss 2004; Barrett, Dunbar & Lycett 2002, Barkow, Cosmides & Tooby 1992). The way these texts present creationism as the *only* possible alternative to evolutionary psychology would make most readers opt for the latter as a matter of logical necessity! Yet it seems to me that the robustness of a theory can only really be tested in relation to its staunchest critics – it is not enough to dismiss these in aggressive ways, in the manner of Stephen Pinker, who describes alternative models of brain functioning that rely instead on the looser idea of the evolution of flexibility, as "no better than magical accounts". (Pinker in Buss 2004 p.56.)

Of course, he is, in part, responding to the somewhat insulting comment by Stephen Gould that evolutionary psychology more often than not is not much more scientific than "Just So Stories", relying on circular arguments (Gould 2000 pp.85-100). Why? In the first place, Gould argues that evolution by natural selection fails with respect to culture because spread of cultures is very different from that of genes. He believes that culture is not illuminated by comparison with genes, for culture is passed from one generation to the next in a labile way that has more in common with Lamarckian than Darwinian ideas. He then evokes his concept of evolutionary "spandrels" – characteristics that are adopted later for other uses than those (if any) any for which they arose, so that it is an error to suggest that those later uses *explain* their existence. For Gould, universal behaviours are such spandrels, and so fall outside any ultra-Darwinian theory.

The biologist Steven Rose also has serious reservation about evolutionary psychology. He believes that a consistent feature is to mistake "enablement" for "causation", so that the framing limitations of physical and chemical processes in the brain do not so much *cause* behaviour, rather, they merely make it possible (Rose 2000 pp.247-65) [2]

---

[2] Critics enthusiastic about evolutionary psychology have suggested that such criticisms sound increasingly 'desperate'. While this collection's merging of accounts of evolutionary psychology with Wilson and Dawkins was somewhat

This objection also applies for religious belief, as I suggested above. In other words, evolutionary psychologists will seek to explain, for example, the supposed universal preference for green by searching back into the era of evolutionary adaptedness (EEA) and speculating about the preferences of hunter-gatherers. Rose makes the point that there are much simpler and more *proximal* accounts of preferences found in contemporary humans, relating to human development, history and culture, than any attempt to find analogies in the earliest period of hominid evolution. He has also argued that living systems are not merely passively sandwiched between the demands of the genes and the challenges of the environment, but *actively construct their environments*, constantly choosing, absorbing, and transforming the world around them. This is very different from viewing behaviour in terms of evolutionary adaptations. While Stephen Mithen's idea of cognitive fluidity is far more convincing than the strictly modular account, it still seems to be premised on a view of behaviour which parallels that of evolutionary psychology, namely that behaviours are products of natural selection , rather than having emerged in other ways.

The Evolution of Religious Consciousness

The specific evolution of religious consciousness is a case in point. I think we can all agree with Mithen's conclusion from paleo-archaeological evidence that the Upper Palaeolithic people are more likely than not to have been the first to believe in supernatural beings and possibly an afterlife (Mithen 1996 p.198). The matter for discussion is whether this evolved from "collapse of barriers that had existed between the multiple intelligences of the Early Human mind". His search for universal features in all religions is a strategy characteristic of all evolutionary psychology, namely, that such universal features point to evolutionary origins. The universal characteristics of a religion are named as: survival of a non-physical component after death, inspiration from supernatural agencies, and ritualistic observances believed to produces change in the world. It would be trivial to cite exceptions to these generalisations. What is more important is the explanation, given by Pascal Boyer and seemingly

---

unfortunate, the seriousness of the criticism of these fellow biologists did not sound to me to be desperate at all.
See also Talbot (2005). Talbot argues for the evolution of paradoxical tendencies in humans. It is not clear how helpful this might be, for if humans behave paradoxically, then almost any behaviour can be accounted for. A simpler explanation is that humans with evolved capacity for flexible responses are drawn to different tendencies depending on social and environmental pressures.

accepted by Mithen, of why these characteristics have emerged: namely, that "the characteristics of supernatural beings as found in religious ideologies relate to the intuitive knowledge about the world genetically encoded in the human mind" (Mithen 1996 p.201). It is the idea of some sort of *genetic* coding, implying that the specific religious instincts defined in the manner above evolved through natural selection, that I find problematic. Drawing on Rose, it seems to me more likely that more *proximal* explanations relating to history and culture favoured the emergence of religious concepts and traditions, rather than evolution through natural selection.

"Two-" and "three-dimensional" mental processes

The assumption in all such cases, of course, is that human intelligence as a whole is something that has been selected for by natural selection. Yet the intelligence-difference one speaks of in such a context is still one of degree, rather than absolute difference, compared with the non-human world. The philosopher Peter Munz has put forward a rather different and equally plausible argument. In his critical engagement with evolutionary psychology he suggests that the evolution of the very large human brain was in fact a liability (Munz 2004). Most other authors have suggested that the increase in brain size arose as a response to social conditions, including the size of social groups. [3] But this leaves unanswered why such social groups emerged in the first place. Others, such as Terence Deacon, argue that the ability to symbolize is related to brain size. Munz contends that this does not adequately distinguish between language that relates to definable objects, and language that relates to events and objects that have not yet happened, and/or are not yet there. The former states meanings about what are, the latter meanings about what are not. Both monkeys and humans can produce symbols, but only humans produce symbols about what is not. In other words, only humans worry about their own uniqueness, and other such problems.

Munz characterises these differences as "two-dimensional" and "three-dimensional" language respectively (Munz 2004 p.44). He suggests that too large a brain does not allow simple and straightforward responses. He also suggests that the very large brain arose in the first instance because of

---

[3] Such authors include, for example, Nicholas Humphrey, Robin Dunbar and Terence Deacon. Of particular note are Humphrey (1976), Dunbar (1996) and Deacon (1997); see also Bryne and Whitten (1988).

a predominantly fishy diet.[4] In this scenario, only those who developed three-dimensional language survived, as it served to compensate for the over-large brain. The large brain was, in this case, a spandrel, rather than something "designed" by natural selection. While the emergence of language also depended on suitable anatomical changes, such as upright posture, and a larynx suitable to modulate and inflect sounds, the different "neural churnings" could come together into a representation. In this sense:

> ... the damage initially caused by too large a brain was not only mitigated, but turned into a positive advantage. With the help of three-dimensional language, it now became possible to assemble linguistically what was taking place in the brain's separate parts. It became possible to assemble the separate cerebral reactions to colour, movement, location, duration and so on into a coherent representation of something which the brain in itself had not taken in as such. The three-dimensional language does not depend for its meaningfulness on ostensive definition (Munz 2004 p.146).

Munz is arguing, then, for the importance of human cultures that made the modification of two-dimensional language into three-dimensional language possible. In this he parts company with the view of evolutionary psychology that specific mental modules, including those for moral and religious sense, have evolved through pressures of natural selection.[5]. The advantage of using language as a means to specify human uniqueness more precisely is that it shows that the difference from the non-human world is cultural, as well as biologically related to the way human minds work. Although there has been a turn towards more relational understandings of *imago Dei* in theological discussion, this scientific discourse is a reminder that human capabilities, such as intelligence, rationality and so on, cannot be separated from the ability to have relationships.[6]. In considering the "morality" of animals I suggest that this

---

[4] He is drawing here on the work of Robin McKie as well as of Finlay, Darlington & Nicastro; for these references see Munz, (2004 p.207).

[5] Space does not permit full discussion of this issue here, but I do share Munz's reservations over the standard evolutionary psychology model and his support for the alternative view of a "general purpose" mind. I discuss this in more detail elsewhere (Deane-Drummond 2009).

[6] Wentzel van Huyssteen (2003) discusses the different theological approaches to image-bearing and their relationship to evolutionary biology, and takes this subject further in his masterly book of 2006. Perhaps somewhat surprisingly he does not seem to challenge the view of evolutionary psychology that the human mind is the way it is because of natural selection, and sees this as simply an extension of

need not reduce or weaken the claim for human uniqueness. In fact it might help to clarify more precisely the difference in the moral landscape of human and non-human animals. For instance, I suggest that we could profitably speak of morality as that which is *three-dimensional* in humans, but only *two-dimensional* in animals.

The awe-struck chimpanzee

I am not, by suggesting this, doubting that the human mind evolved in a way that enabled both morality and religion to emerge. Current comparison between human and non-human species gives important insights in this respect. I also warm to the possibility that non-human primates may have a theory of mind, and have no problems with blurring the boundary between human and animal, with all that this might imply. For example, there is also the possibility that non-human primates feel a sense of wonder and this capacity, like shame, reflects a biological propensity for a religious sensibility. Marc Bekoff is among those who have spent many years observing animals, and he believes that feelings akin to wonder do exist among them. Thus he writes:

> Sometimes a chimpanzee will dance at a waterfall with total abandon. Why? The actions are deliberate but obscure. Could it be they are a joyous response to being alive, or even an expression of the chimp's awe of nature? Where after all, might human spiritual impulses originate? Jane Goodall wonders whether these dances are indicative of religious behaviour, precursors of religious ritual. "If the chimpanzee could share his

---

evolutionary epistemology in becoming "embodied". Thus he speaks of forms of human intelligence as "adaptations" in a way that would be quite foreign to someone of Peter Munz's perspective. While he warms to Popper's view of knowledge as that which is created by human intelligence, with the least successful theories eliminated, this version of evolutionary epistemology is very different from more biologically-based theories such as those of Henry Plotkin (1994). Van Huyssteen even speaks approvingly of a *universal Darwinism*, which is odd, considering his discourse with postmodern theory. The crucial difference is that between a general-purpose mind and an adapted mind, and van Huyssteeen seems to lean towards the latter (see van Huyssteen 2006 pp.75-83). Munz has pointed to the underlying positivism in evolutionary psychology, that ties the mind into specific instructions, coded by genes, rather than seeing the mind as freely inventing hypotheses. Munz is therefore sharply critical of authors such as Henry Plotkin, whom van Huyssteen seems to endorse, and views the path that Popper took in his incorporation of Darwin as very different from the ultra-Darwinian stream of evolutionary psychology (see Munz 2004 pp.156, 165-6).

feelings and questions with others, might these wild elemental displays become ritualised into some form of animistic religion? Would they worship the falls, the deluge from the sky, the thunder and lightning – the gods of the elements? So all-powerful; so incomprehensible" (Bekoff, 2007 pp.61-2).

Steven Mithen describes our capacity to think of animals as like ourselves as anthropomorphism, and believes that this emerges directly as a result of the cognitive fluidity of the human mind. He comments that:

> … we seem unable to help anthropomorphising animals – some claim it is built into us by nature and nurture – and while this gives us considerable pleasure, it is a problem that plagues the study of animal behaviour, for it is unlikely that animals really do have human-like minds. Anthropomorphism is a seamless integration between social and natural history intelligence (Mithen 1996 p.188).

And it was just this kind of intelligence that was missing in our Neanderthal relatives, who, like our mutual primate relatives, had "isolated domains" in their mental intelligence. Nonetheless, Mithen speculates that Neanderthals did have a more advanced form of communication compared with primates, using a "holistic" form of communication that was more like singing, providing the root for not just language, but also music – and arguably, though Mithen does not say as much, religious liturgy as well. Yet this is all dependent on how we understand the analogies we find in other species. I am quite certain that the wonder that chimpanzees feel bears only a very loose resemblance to the capacity for the human spirit to wonder in face of nature, or even a religious awe that arguably goes further in terms of experiential depth, to Rudolph Otto's sense of the "Holy Other". Yet that does not mean that the analogy is unhelpful, for it is the closest we can get to trying to understand what it might be like to be another creature.

Finally ….

Finally, I doubt that evolutionary psychology is guilty of the kind of "nativism" and allegiance to misconceptions about human nature that plagued Nazi Germany, as some extreme critics have suggested. Such comparisons are unhelpful. But perhaps our historical memory serves once more as a reminder that ideas are never isolated from the way that they have been used in the past. While evolutionary psychologists will cringe at such a comparison, it is easy to see why it is not just feminists who find some of the explanations of current human behaviour in the terms of evolutionary psychology objectionable, for they appear somehow to soften any sense of guilt about particular tendencies and practices – as evolved tendencies, they seem like almost inevitable outcomes, even if those who have developed the field try to circumvent this problem by arguing for "decision rules" that give an element of flexibility in behaviour (Buss 2000 p.56).

Mithen's idea of cognitive fluidity is important in as much as it softens the "Swiss army knife" position of other evolutionary psychologists. Yet his idea creates in its turn a tension that is not fully resolved, namely, how far are religious instincts properly considered to be caused in their details by evolved capacities, and how far merely *enabled* through a cognitive fluidity, drawing on historical and cultural learning that was also taking place? I am also hardly surprised that archaeological evidence has failed to distinguish the two possibilities that Mithen names at the start of his paper; one would not expect a sudden revelation of God to appear like a bolt out of the blue, but rather, revelation to come when humanity was ready to receive it. Whether one chooses to believe or not to believe cannot be confirmed or discredited by archaeological science. The most sensible view, therefore, is not to claim to have found either a basis for natural theology or a basis for atheism embedded in the archaeological record. Rather, the record will give us pause to reflect on human uniqueness, in its capacity for religious knowledge, and at the same time on human commonality with other species.

References

Barkow, J.H., Cosmides, L & Tooby, J. 1992. *The Adapted Mind.* New York: Oxford University Press.

Barrett, J.L. 2004. *Why Would Anyone Believe in God?* Lanham/Plymouth: Altamira Press.

Barrett, L., Dunbar, R. & Lycett, J. 2002. *Human Evolutionary Psychology*. Basingstoke: Palgrave MacMillan..

Bekoff, M. 2007. *The Emotional Lives of Animals* Novato: New World Library.

Bryne, R. & Whiten, A. (eds.) 1988. *Machiavellian Intelligence: Social Expertise and the Evolution of Intellect in Monkeys, Apes and Humans.* Oxford: Clarendon Press.

Buss, D.M. 2004. *Evolutionary Psychology: The New Science of the Mind,* 2nd *edn* Boston: Pearson Education.

Deacon, T. 1997. *The Symbolic Species.* London: Allen Lane, Penguin Press.

Deane-Drummond, C. *Christ and Evolution.* Minneapolis: Fortress Press, due 2009.

Dunbar, R. I. M. 1996. Determinates of Group Size in Primates: A General Model', in W.G. Runciman, J.Maynard Smith & R.J. M. Dunbar (eds) *Evolution of Social Behaviour Patterns in Primates and Man.* Oxford: University Press.

Gould, S.J. 2000. More Things in Heaven and Earth, in H. Rose & S. Rose (eds), *Alas Poor Darwin: Arguments Against Evolutionary Psychology* London: Jonathan Cape.

Humphrey, N. K. 1976. The Social Function of Intellect, in P.P.G. Bateson & R.A. Hinde (eds) *Growing Pains in Ethology.* Cambridge: University Press.

Mithen, S. 1996. *The Prehistory of the Mind* London: Phoenix.

Munz, P. 2004. *Beyond Wittgenstein's Poker: New Light on Popper and Wittgenstein* Aldershot: Ashgate.

Plotkin, H. 1994. *Darwin Machines and the Nature of Knowledge.* London: Allen Lane. Republished Penguin Books, 1995.

Rose, S. 2000. 'Escaping Evolutionary Psychology', in H. Rose and S. Rose (eds), *Alas Poor Darwin: Arguments Against Evolutionary Psychology* London: Jonathan Cape.

Russell, R.J., Stoeger W.R & Ayala, F.J. 1998. *Evolutionary and Molecular Biology: Scientific Perspectives on Divine Action.* Vatican City/Berkeley: Vatican Observatory Publications/Centre for Theology and the Natural Sciences.

Talbot, C. 2005. *The Paradoxical Primate* Exeter: Imprint Academic.

van Huyssteen, W. 2003. Fallen Angels or Rising Beasts? Theological Perspectives on Human Uniqueness, *Theology and Science*, 1 (2) pp.161-78.

—. *Alone in All the World? Human Uniqueness in Science and Theology* Grand Rapids/Cambridge: Eerdmans.

CHAPTER THREE

DIVERGING PATHS IN THE STUDY
OF THE EVOLUTION OF RELIGION

LLUIS OVIEDO

Lluis Oviedo Torró OFM, although himself Spanish, is Professor of Theological Anthropology at the Pontifical University "Antonianum", Rome. Author of the books: "Altruismo y caridad" (Altruism and charity 1998); "La fe cristiana ante los nuevos desafios sociales" (Christian faith and the new social challenges 2002), and many scholarly articles on issues of religious faith, science and society.

This essay derives from a submitted contribution to the 2007 conference, but it seems most appropriately placed at this point in the printed volume. Prof Oviedo argues that an approach broader not only than Stephen Mithen's, but even than Celia Deane-Drummond's, is necessary if we are to do full justice to the socio-cultural role of religion – and that only by giving this role its due weight, as supplementing though not supplanting biological mechanisms, can we hope to formulate a full perspective on the factors leading to the evolution of religions.

A diversity of approaches

In the last few years religious evolution has become a hot topic, as increasing numbers of scholars from different backgrounds try to figure out how religion could have arisen at all in the human mind, and how it developed, becoming a strong part of the human psyche, of communication, and of social organization.

Evolution is a strongly "biological" issue; it is in relation to biology that this concept is most at home. So it is to be expected that a biological approach will provide the most appropriate tools and insights to show the intimate dynamics of religious origins and development. As Daniel

Dennett has stated, evolution becomes a "universal acid... it eats through just about every traditional concept and leaves in its wake a revolutionized world" (Dennett 1995 p.63). Religion will not prove an exception, and the acid is just now performing its function of altering the traditional ways in which religion has been understood, thereby trying to achieve its revolutionary promise. The crux of the argument resides in the acknowledgedly counter-adaptive traits resulting from certain religious ideas and activities. As some have put it: if one invests more time in learning fighting and hunting skills and less in praying or invoking deities, one's chance of survival will improve (Mithen 1996).

Despite this apparent limitation, there are even several alternative biological accounts of religious evolution. I will introduce the main current theories in the next section. Such pluralism is not surprising, if we take into account the many survival strategies existing in life, aimed at enhancing the chances of reproduction and fitness. Religion can easily be assimilated into the same dynamics, as can all human and social realities which have helped people to overcome the harshness of their respective environments.

Despite the abundance of proposed explanations for this evolutionary process, it is worth asking whether this is the only, or even the most helpful approach in modelling the evolution of religion; whether it is complete (not requiring further explanations); and whether alternative paths could improve the explanatory power of our theory, complementing an outlook that is otherwise too one-sided and reductionist.

At the risk of stating the obvious, Dennett is right that evolution may be understood as a "universal acid" in the sense that all historical realities must pass through by the process of variation, selection and renewed stabilization. The hard question is whether the factors governing this process are just the ones described by the biological-adaptationist approach (survival and reproductive advantage), or whether different human and social realities are governed by factors or strategies that are not directly linked to biological adaptation.

Bearing this in mind, we can surely follow some alternative paths to enrich the picture of religious evolution, resorting, for example, to the role played by symbols, their own evolution, and the development of social forms linked to religious ideas – that is to say, all that we could put under the broad umbrella of "co-evolution" between genetic and cultural variables. It seems appropriate to look for the specific logic presiding over the evolution of religious ideas and language. Trying to go further than the simple adaptationist programme, it is useful to place religion within a distinctive evolutionary framework, where forces beyond sheer biological adaptation come into play. The recent discussion, promoted by Jerry Fodor (2007) among others, stressing the internal factors "channelling" evolution, rather than mere adaptation to external pressures, should be taken seriously into account, if we wish to overcome the shortcomings of many alternative propositions.

1. The biological evolution of religion, as studied so far

James Dow has recently offered a good classifiction of the main exisiting theories on the evolution of religion. He lists three kinds:
- "commitment theory", which postulates that religion is a costly system of signaling that reduces deception and creates cooperation within groups
- "cognitive theory", which postulates that religion is the manifestation of mental modules that have evolved for other purposes; and ...
- "ecological regulation theory", which postulates that religion is a master control system regulating the interaction of human groups with their environments (Dow 2006 p.67).

Dow attributes the three theories respectively to Richard Sosis (Sosis & Alcorta 2003); Scot Atran and Ara Norenzayan (2004); and Roy Rappaport (1999). The first two theories appear to represent the positions of the parties involved in a recent anthropological discussion between so called "adaptationists" on the one hand and "by-product theorists" on the other – in other words, between those who attribute to religion some adaptive function helping to improve the survival chances of a population, and those who reject any useful contribution of religion, preferring to treat it as a derivate of the hyperactivity of some mental processors. Dow suggests an integration of genetic components with cultural ones in order to provide a more comprehensive understanding of the evolution of religion. At the same time he rejects the ideas of David Sloan Wilson (2002) concerning "group selection" as an explanation for the adaptive power of some religious forms, i.e. those which help to improve cohesion and collaboration within a group.

There are some more theories, which can be suggested to support the case for a biological evolution, or at least to complete the typology provided by Dow. Examples are provided by recent theories on "natural selection of cultural variations" (Richerson 2005 pp.76 ff.; Wexler 2006). However, in all these cases the authors deal with a combination of sheer evolutionary theory and the role played by cultural inputs, which can be perceived at work both at the ontogenetic level (through brain development during children's growth) and at the phylogenetic (succession-of-generations) level, where "mental representations and public productions alternate" and are governed by a "selection model" (Sperber 1996 p.99).

Some explanations of phylogenetic selection take into account the cladistic[1] mechanisms governing all biological evolution, and show how religion evolves along similar paths to those described by cladistic or phylogenetic trees:

> In some cases, religious innovations were subjected to cultural selection and either failed or experienced successful propagation, depending on the power of the cultural, selective factors ... Judaism, Christianity and Islam are cladistically related and all started as horizontally transmitted[2] innovations (Stone & Lurquin 2007 pp.238 ff.)

Other researchers have proposed alternative ways of characterising the evolution of religion. One of these is the neurological view of religion as the result of an "interplay between the lower and the upper brain". Thus:

> Religion may be partly the product of humanity's intuitions of its dual interiority and the fruitful creative Spirit generated by the interplay of the gene pool, as in "the Ancient of Days", and the upper brain, as "Logos".
> (Turner 1986, 176).

Last, we should register the attempts to apply thermodynamics to the evolution of religion, as the most basic force governing every natural process, not only in the physical realm. Thermodynamics refers to a sparing use of energy, the pursuit of order, and the practice of self-organization; all these may actually be applied to religion and its evolution in a quite obvious way, beside the laws of "survival of the fittest" (Dean 2004).

It is easy to perceive an "evolution" in the evolutionary study of religion! The last few years have seen a move from a rather rigid biological stance to an outlook which tries to take into account cultural elements or dimensions, giving rise to a more complex picture of the evolution. In my view the process is still incomplete, and will remain so until we take into account other variables involved in religious evolution, especially at the historic level, involving changes in religious symbols and ritual practices.

---

[1] Cladistic = sharing features indicating common ancestry. Ed.
[2] Horizontal transmission = spread among members of a group, living at the same time. Contrast "vertical" = spread from generation to generation. Ed.

2. Alternative paths in studying the evolution of religion

Probably the anthropological questions should be asked in a more radical way if the biological understanding of religion is to be challenged and completed. Only a very narrow picture can assume that human beings are merely guided by the laws of survival and reproductive fitness. There are several other conditions a person needs to fulfil in order to achieve satisfaction in life: without them many deem life devoid of meaning or purpose, even when they enjoy all the resources which ensure survival and all the means of achieving their reproductive goals.

At this point it is worth recalling some of the limitations of the biological approach to religious evolution:
- disregard of the more conscious dimension of the religious experience,
- dismissal of the role of rationality in the development of religion through history,
- neglect of the role played by feelings and emotions in religious processes,
- disregard of the role played by symbols and ritual activities,
- neglect of the communicative interplay linked to every religious activity.

In my opinion, a more complex theory of religious evolution should take into account at least these neglected dimensions, and should broaden the evolutionary criteria to include these additional aspects of human make-up. In brief: the personal needs which are not strictly confined to biological or basic requirements, are nevertheless to be satisfied following a similar logic to the one presiding over biological evolution, which, by the same token, should be considered in terms reaching beyond strict adaptationist orthodoxy.

It is possible to reconstruct quite concisely some paths that religious ideas and forms have followed in fulfilling the human needs we have just spelled out.

a) Linking consciousness to religious experience

Religion may be seen as a way to implement certain levels of consciousness. If it is acknowledged that consciousness plays a central role in defining human personality, albeit through rather complex processes (Donald 2001), religious ideas and rituals may be conceived as ways to implement such a faculty, giving rise to levels of awareness which, at the same time, could improve communication and sociality, and deepen knowledge of one's own life and environment. In this sense, religious ideas could evolve while helping to enhance such levels of awareness. Some historic religions could be examined in these terms. The example of Covenant Judaism is paradigmatic of a religious form which helped to raise the awareness of the follower's own self and his/her sense of responsibility, and gave a new configuration to social relationships. In the Christian context, St. Augustine's exploration of his own consciousness opens up a new understanding of the human being and its greatness, as this self-awareness becomes the place of God's revelation. It is arguable that these "improvements" also generate some kind of "biological return" (perhaps by improving levels of fitness?). Nevertheless this evolution is more closely linked to the fulfilment of other types of human need. Only as a subsequent development may it indirectly help to improve survival and reproductive chances; even then the religious aspect is sometimes not responsible for such improvements, and may even challenge the biological needs. This appears to underline the peculiarity of human nature in the biological realm.

b) Religion and rationality

Religion evolves following requisites of rationality. This is a classic point, suggested in the most explicit way by Max Weber, and further unfolded in the works of Niklas Luhmann (1977) and Lawrence Iannaccone. The latter has applied "rational choice theory" to explain many religious activities (Iannaccone 1998). In brief, he argues that religion has evolved to help us come to terms with the increasing requirements of an ever-more-rational mind, or a mind that could not accept former descriptions and ideas about the divinity and its behaviour, and was looking for more congenial explanations. Rationality is a rather multi-faceted concept, and is understood in quite different ways along philosophical and sociological lines; in broad terms, it addresses a human search for argumentative clarity, convincing ideas, and the best way to cope with one's own challenges. It is clear that the development of greater logical skills and the use of this ability to deal with one's personal and social life exerts a pressure on religious ideas, making them an integral part of the human mindset and one of the resources suited to tackling certain problems. As Weber stated, religions evolve following a pattern of "rationalization" in the attempt to adapt to human and social needs and to provide better models for guiding action.

It is quite easy to follow this pattern in some canonical religious texts; for example, in the Old Testament, we see how the wise men of Israel try to rethink eschatological expectations in a way that can satisfy some rational demands. The same happens in Christian history with some of the reforms the church has experienced. As Weber showed (1976 [1906]), when these reforms succeed, they help to deal with new rational and practical requirements. This process also needs to be linked to the ethical demands of society, which should be respectful of rational considerations, but require frequent religious support. The argument concerning rationalization may be followed in other authors; some of the views suggest that the interplay between rationality and religion is a rather complex one, and religion is not a simple dependent variable of the rationalization process, but rather it triggers a more rational mindset and social organization. In any case, the history of religion may be reconstructed along these lines, in a complementary role to pure biological needs.

c) Religion is concerned with emotion

Religion evolves as a means of coping with emotional necessities, or as an instance of "emotional management". The idea is similar to what has been stated in the above points: human beings have emotional needs and must be able to cope with emotional distress; as a result, religion may be seen as a powerful instrument helping sometimes to channel emotions, and sometimes to tame them. Religious rituals are perhaps the most customary instruments to perform that task. It is reasonable to think that religious forms evolved better to cope with such needs, or in order better to use that potential, or to tame it for social purposes. A particular case is presented by the problem of guilt; this requires appropriate management, along sacrificial or other lines. Everything points to the idea that religion has evolved to deal with this negative – but functional – emotion. On the other hand, it is useful to recall that this evolution has been accompanied by a development of artistic forms: visual, dramatic and, especially, musical. It is impossible to imagine a religious expression or ritual without artistic enrichment, which usually does not relate to biological needs, but rather to the way humans live and experience reality.

d) Religion is a symbolic expression

Symbols are a central part of every human evolution. As Terrence Deacon, one of a long line of "externalists", has highlighted in the last decade, we cannot understand the human evolutionary process unless we take into account the central role played by symbols and language (Deacon 1997). Symbols contribute to the second pole of the co-evolutionary process, involving genetics and culture. If this is true for the general evolution of humankind, it should also be true for the evolution of religion, as religion unavoidably resorts to symbols, refines them, and expands them through history. Any project attempting to model this process should try to explain how religious symbols change and adapt to new situations and requirements. Recent theories resort to blendings, distinctions, and drifts of meaning, among other means, to figure out what is going on. Surely more than this is involved, even though the natural logic underlying the process could be similar to the biological one: selection, from among many variations, of those symbols or meanings most suited to coping with anthropological, social, and specifically religious needs. However, symbols represent a clear case of "evolutionary channelling", as their adoption very often determines the present and future of a human group.

e) Religion is a form of communication

Communication is a central issue in religious experience, and complementary to symbolic ability. It is right to think that religion has evolved to satisfy communicative necessities in a population in which this capacity was crucial both for improving survival chances and for other aspects of development. There are some available analyses dealing with the ability of religious ideas to be transmitted despite their counter-intuitive character. I suggest a different path by looking at religious evolution as a way of improving communication among group members, by providing new ideas and concepts, or by enriching the available pool of categories for comprehending and describing reality. Another way of viewing the same issue is to consider religious communication as a resource aimed at reducing uncertainty, or, as other scholars put it, at allaying anxiety. In any case, religion and its evolution should be linked to the expansion of communicative skills in society, a process that further stresses the rather "channelling" character of human evolution, strongly tied in with specific anthropological and social traits.

To conclude

There are certainly several more options for mapping the evolutionary path of religion than the crudely biological one. My suggestion assumes a more complex anthropology in which human beings experience a range of needs, beyond sheer survival and reproduction. In this sense, the contribution of culture in the process of "co-evolution" is not seen as merely improving survival and reproductive chances, but as a means of providing a better "quality of life" as advanced humans experience specific needs that can only be satisfied by cultural means. Humans at this stage of evolution – in contrast with other animals – do not simply need to survive and mate, but also to seek a life worth living. Religion is clearly more closely linked to this second set of factors, than to those concerning biological fitness.

Finally, we should ask whether religion might evolve purely following its own "religious needs", perhaps distilling all the aforementioned factors into a single mindset. Religion pursues its own logic, once it has reached a certain "critical mass" in the evolutionary process. In other words, it follows its own pattern, reflecting the religious needs of a population. In this sense, religion becomes part of the human equipment that "channels" the evolution of the species in an intrinsic way, beyond or beneath the adaptive pressures.

A further task would be to describe and organize the different types of religion that reach a relative stability after a long period of development. I am convinced that religion evolves not in a linear way, but rather like a bush, in which some branches or *clades* coexist, and find natural niches at the same time and in the same social and cultural places. This task is not so difficult as the one undertaken by some projects, at least since the typology of Robert Bellah[3] (1991). I am convinced that some new typologies should be taken into account if we wish to figure out how religion stabilizes, or around which "attractors" religious forms tend to converge.

An optimal starting point could be the proposals of some cognitive anthropologists, like Thomas Lawson and Robert McCauley (1990), with their work on ritual, or Harvey Whitehouse (2004) with his distinction between *imagistic* and *doctrinal* religion. It is my opinion that these distinctions should be enriched to encompass, for example, mystical and moral religious forms. Furthermore, some anthropological ideas from former generations of anthropologists could help in this endeavour, as is the case with Mary Douglas's typology, based on Basil Bernstein's more

---

[3] Bellah proposes a taxonomy of religion's evolution in five stages: primitive (e.g., Australian Aborigines), archaic (e.g., Native American), historic (e.g., ancient Judaism, Confucianism, Buddhism, Islam, early Palestinian Christianity), early modern (e.g., Protestant Christianity), and modern (religious individualism).

positional and more *expressively personal* cultural models, which clearly apply to religious forms (Douglas 2003 p.37 f.).

We are still some way away from a general theory of religious evolution, and the points it tends to establish. I hope some further advances can soon be made towards its foundation.

References

Atran, S. & Norenzayan, A. 2004. Religion's Evolutionary Landscape: Counterintuition, Commitment, Compassion, Communion. *Behavioral and Brain Sciences* 27(6), 713-730.

Bellah, R. 1991 [1961]. *Religious Evolution.* New York: Irvington Publishers.

Deacon, T.W. 1997. *The Symbolic Species: The Co-evolution of Language and the Human Brain.* New York: Norton.

Dean, M. 2008. Cognitive Thermodynamics in Culture & Religion, paper delivered at the Annual Meeting of SSSR, Kansas City, 2004.

Dennett, D.C. 1995. *Darwin's Dangerous Idea: Evolution and the Meanings of Life,* New York: Simon & Schuster.

Donald, M. 2001. *A Mind so Rare: The Evolution of Human Consciousness.* New York: Norton.

—. (2003), *Natural Symbols.* London: Routledge [first published 1970].

Dow, J.W. 2006. The Evolution of Religion: Three Anthropological Approaches, in *Method & Theory in the Study of Religion,* 18, 67-91.

Fodor J. 2007. Why Pigs Don't Have Wings, *London Review of Books* (18 Oct.) 29-20.

Iannaccone, L. 1998. Introduction to the Economics of Religion, *Journal of Economic Literature,* 36, 1465-1496.

Lawson E.T. & McCauley R.N. 1990. *Rethinking Religion: Connecting Cognition and Culture.* Cambridge: University Press

Luhmann N. 1977. *Funktion der Religion,* Frankfurt a.M.: Suhrkamp.

Mithen S.J. 1999. *The Prehistory of the Mind: A Search for the Origins of Art, Religion and Science,* London: Thames and Hudson.

Rappaport R.A. 1999. *Ritual and Religion in the Making of Humanity,* Cambridge, UK - New York: Cambridge University Press.

Richerson P.J. 2005. *Not by Genes Alone: How Culture Transformed Human Evolution,* Chicago: University Press

Sosis, R. & Alcorta C. 2003. Signaling, Solidarity, and the Sacred: The Evolution of Religious Behavior, *Evolutionary Anthropology* 12, 264-274.

Sperber D. 1996. *Explaining Culture: A Naturalistic Approach*, Oxford: Blackwell.
Stone L., Lurquin, P.F. 2007. *Genes, Culture and Human Evolution: A Synthesis*, Oxford: Blackwell.
Turner V.W. 1986. *The Anthropology of Performance*, New York, PAJ Publishers
Weber, M. 1976 [1906]. *The Protestant ethic and the spirit of capitalism*, London: Allen & Unwin.
Wexler, B.E. 2006. *Brain and Culture: Neurobiology, Ideology, and Social Change*, Cambridge, MA: MIT Press
Whitehouse, H. 2004. *Modes of Religiosity: A Cognitive Theory of Religious Transmission*, Walnut Creek, NY: Altamira Press
Wilson, D.S. 2002 *Darwin's Cathedral: Evolution, Religion and the Nature of Society*. Chicago: University Press.

CHAPTER FOUR

DARWIN'S GIFTS TO THEOLOGY[1]

FRASER WATTS

The Revd Dr Fraser Watts is Reader in Theology and Science at the University of Cambridge, Vice-President of the International Society for Science and Religion, and Vicar-Chaplain of St Edward, King and Martyr, Cambridge. He trained first as a psychologist, and worked for a number of years on cognitive aspects of emotional disorders at the MRC Applied Psychology Unit, London; during part of this time he was President of the British Psychological Society. His books include: "The Psychology of Religious Knowing" (co-author, 1988); "Christians and Bioethics" (2000) and "Psychology and Theology" (2002).

Here he follows in the footsteps of Aubrey Moore and Arthur Peacocke by arguing that Darwinian thought, far from being anti-religious, is deeply compatible with, and indeed encourages, the sense of an ever-sustaining God –a God who underlies the evolution of natural forms, not according to pre-ordained patterns but by the operation of divinely initiated processes. Many implications of this position are explored, from the relation of humans to other animals to the position of Christ in the evolutionary story.

At first glance, Darwin's theory of evolution by natural selection seemed to most people to be an enemy of Christianity. However, following in the footsteps of Aubrey Moore, I want to argue that, though evolutionary theory appeared in the guise of an enemy, it has actually proved to be a good friend of Christianity. Christian doctrine has been

---

[1] This paper incorporates material from a sermon on *Christianity After Darwin* given at St Edward's Church, Cambridge on 6th August 2006, and also material previously published in a paper on "Evolution, Human Nature & Christianity", *Epworth Review* 29 (i), 24-31 (2002).

rethought in the light of Darwin and, I believe, at every point it has been rethought for the better. I will argue this in relation to our views on the Bible, on creation, progress and providence, on human nature, and on salvation history. I will also discuss how Darwinism has provided a new approach to understanding the significance of religion.

Bible

Some people saw Darwin's theory of evolution as contradicting the Bible. However, I suggest that Darwin has helped us to see that the first two chapters of Genesis were not meant to be a scientific textbook. They are answering different kinds of questions, so there is no necessary conflict between evolutionary biology and the book of Genesis. More generally, Darwin has helped people to be more conscious of the diversity of material in the Bible, and of the different kinds of truth that it represents.

The book of Genesis, as a whole, is concerned with the origin of the people of God. What is says about the origin of the created order is hardly more than a preliminary to that. Furthermore, it is anachronistic to imagine that the first chapters of Genesis were intended as a literal account of the origin of the world and of species. Rather, they are making theological points about the dependence of the created order on God, albeit making them in narrative form. One reason for saying this is that the first two chapters of Genesis tell different stories about how human beings were created. If the authors of Genesis had meant them to be taken literally, they would have worried about the contradiction. They were presumably happy to include both stories because they both make the same basic point that human beings owe their existence to God.

Some people imagine that taking parts of Genesis analogically is new-fangled liberalism, but that is a misreading of the history of Biblical interpretation. Indeed, it is the insistence found in some contemporary quarters on the literal truth of the scriptures that is new-fangled. It arose in the early part of the twentieth century, largely in America, and as a response to Darwinism. Darwinism did not overturn prior fundamentalist assumptions. On the contrary, fundamentalism arose as a reaction to Darwinism. Awareness of the need to take some parts of Genesis analogically goes back at least as far as St. Augustine, who understood very clearly that if Christians insisted on taking every Biblical detail literally they would make a laughing stock of themselves, and bring the essentials of the Christian faith into disrepute. Darwin has helped us to see that Genesis is not competing with modern scientific theories but making important religious points of its own.

It is significant that the so-called "higher" Biblical criticism, initially developed largely in Germany, was just making its way into the English speaking world at about the time that *The Origin of Species* was published. There was thus a vigorous debate about the authority and interpretation of Scripture, and the Darwinian controversies became intertwined with this debate. It would be a misrepresentation to suggest that the new approach challenged a settled and considered literal approach to biblical interpretation. Rather, there was a growing awareness, in some ways uncomfortable, that greater sophistication was needed about biblical interpretation than had generally been found hitherto. Probably there would have been a move toward such greater sophistication anyway, but it is arguable that it was facilitated by Darwin.

Creation, Progress and Providence

One of the key issues for a reconciliation of evolutionary biology and the doctrine of creation is whether there is scope for evolution to be guided according to God's purposes. Many, such as Jacques Monod (1972), have argued that evolution is a process of blind chance, and as such is incompatible with the view that it represents the purposes of God. However, it is an exaggeration to suggest that evolution is pure chance. Most people assume that mutations are a chance matter, even though the rate of mutation may be affected by environmental variables. However, natural selection is clearly not a chance process, it systematically favours adaptiveness to the environment, and good processing of information about the environment.

The idea that there has been some kind of evolutionary progress is notoriously controversial (see Barlow 1994). However, following Ayala (1998), I suggest that at least there has been "directional change" in evolution, whether that is seen as a movement towards greater complexity, or (better) towards greater capacity for processing of information. Though some evolutionary theorists want to deny directionality, it seems to me difficult to do so. Evolution clearly represents a general movement towards more complex forms, and greater capacity for processing information. Evolution is not completely random and *a*directional, nor does it represent a gradual deterioration in adaptiveness.

This is not to claim that the fossil record tells a simple story; in fact there are many twists, turns and blind alleys. Neither is it to claim that it was inevitable that evolution should take exactly the course it did; the movement towards more complex creatures with greater capacity for information processing could clearly have come about in many different ways. Finally, it is not to claim that the culmination of this evolutionary process had to be *Homo sapiens*; it could equally have led to complex creatures, good at information processing, that were different in detail from ourselves.

The directional change of evolution should not be confused with the idea that the evolution has, in every way, made things better. Human beings have a greater capacity for both good and evil than other species, and evolutionary developments usually seem to be morally ambiguous. I suggest that it is sufficient for the theological interpretation of evolution to maintain that there has been directional change consistent with God's purposes. Early theological appropriations of Darwinism owed too much to general assumptions about progress that have become very suspect. However, I suggest that progress is unnecessary and that directional change is enough (Peacocke 2001). For a Christian, it is highly significant that evolution has led to creatures that are capable of receiving God's revelation of himself, becoming aware of his presence in creation, worshipping him and seeking to fulfil his purposes. Indeed, Christians now mostly assume that it was God's purpose that such creatures should evolve by some route or other.

The question remains of what process led to such creatures. That is a scientific question, as yet not fully resolved, and close to the heart of the current debate between narrow and broad evolutionary theorists. An increasing number of such theorists are concluding that there may be other processes at work apart from natural selection. Indeed, there is a gulf opening up between the narrow evolutionary theorists who admit no other processes apart from natural selection, and the broad evolutionary theorists who do allow other processes. There is thus a debate about whether we already understand all the essentials of evolution, or whether there are many unexplained things about evolution that we do not yet understand.

The position of Simon Conway Morris on this is very interesting (Conway Morris 1998). He dissociates himself from the view that evolution is a matter of pure chance. Equally, he does not accept the view that evolution is the product of natural selection alone. On the other hand, he obviously does not want to be associated with "intelligent design". It is interesting to see a major evolutionary theorist, with Christian sympathies, developing a theoretical approach that can be distinguished from any of those positions. It is one in which evolutionary convergence plays a significant role. More radical still is the approach of Stuart Kauffman (1995) who posits a tendency for self-development towards complexity. Though the jury is out on this debate, the narrow theorists should not be allowed to get away with claiming that theirs is the only possible or scientifically respectable position. It is not even clear that they have Darwin himself on their side.

Darwin has helped us to recover the sense that God's creative work is continuous. In Darwin's time there were many "deists" who took the heretical view that, though God had created things once for all at the beginning, he was no longer actively involved in creation. Darwin helped people to understand that creation is an ongoing project for God. That is actually the point that Aubrey Moore had in mind when he first said that Darwin appeared in the guise of an enemy, but proved to be a friend of Christianity; he meant that Darwin had helped Christians to recover a stronger sense of the continuing nature of God's creative work (Moore, J.R. 1979). Creation is not something God did once for all. Rather, we now see that evolution has been God's on-going way of creating species. Evolution is a remarkable way for God to have brought species into being.

From the outset, some Christians have seen the possibility of reframing natural theology in the light of evolutionary biology. Charles Kingsley (1819-1875) was one of the early advocates of this approach. In his view it was a "loftier thought" that God should have created primal forms capable of further development than that he should have created every species fully formed (Brooke 1991). Darwinism thus led to an enhanced wonder at God's creative powers. Of course, this revised natural theology was, in some sense, weaker than the old one which had reached its culminating expression in the writing of William Paley (1802). It did not attempt to argue for the existence of God on the grounds that there could be no other explanation of the fruitfulness and adaptiveness of the natural order. It simply adopted a theological view of things that could be set alongside the scientific one as a complementary perspective.

Human Nature

Darwin also helped to undermine an over-exalted view of human nature. Human beings sometimes like to believe that they are the pinnacle of creation, and the Victorian world had an exalted and unbalanced view of human dignity, which was threatened by Darwinism. People thought that if humans were descended from monkeys, Darwin was saying that humans were just animals, and they found that very offensive. It is sometimes imagined that an exalted view of human dignity is a central tenet of Christianity, but it really isn't. We have here an example of a recurrent phenomenon in the history of the relationship between theology and secular thought. It frequently happens that Christian belief becomes so closely intertwined with certain assumptions that are not themselves specifically Christian that, when those assumptions are challenged, people assume that Christianity itself is being challenged. The assumptions about human dignity that were challenged by Darwinism were never specifically Christian.

Darwin has helped us to recover a more balanced view of how good and evil are intertwined in human beings. We humans have evolved to the point where we have a unique capacity to do things deliberately. The struggle between good and evil has been brought to a new height in human beings, because we are more knowing than any other species. Thus Darwin has helped us to see that humans are both better and worse than other species.

There is a further problem of how far human nature itself can be understood in terms of its evolutionary origins, which raises the vexed question of "reductionism". There are several strong reductionisms current in the human sciences, and it is a feature of them that they come with phrases like "nothing but...", or "no more than..." (Watts 2002). For example, it is clear that the human mind functions in some ways like a computer, but is it "nothing but" a computer program? It is also clear that the physical brain is involved in all human experience, but is such experience "nothing but" the product of the physical brain? It is equally clear that much human experience is shaped and formed by culture and language, but is it "nothing but" the product of culture? Evolutionary reductionism runs parallel to these other cases. Granted that human beings are the products of an evolutionary process of some kind, are they "nothing but" the products of evolution?

These reductionisms trade on an exaggeration of analogies, on a tendency to claim an exclusive position, and on illegitimate inferences. For example, there is a useful analogy between the human mind and a computer, but the closeness of the analogy can be exaggerated. All of these reductionisms identify helpful approaches to understanding human nature, but none of them is the only useful approach; none should be given an exclusive significance. It seems to be a feature of something as complex and multi-faceted as human nature that it requires multiple approaches to understand it. Further, you cannot move legitimately from an explanation of how human beings have arisen to what they are really like. Knowing that human beings have evolved does not allow you to conclude that they are nothing more than products of evolution. That would be to commit the "genetic fallacy".

The debate about the evolutionary explanation of human beings has focused as much on morality as anything else (see Ruse 1995, Thompson 1995). However, there are various ambiguities as to what is being claimed in "evolutionary ethics". It tends to be assumed that moral behaviour in human beings is essentially similar to the capacity of other animals to sacrifice themselves in the interests of their kin. Certainly, human moral behaviours have their antecedents in such animal behaviours, but it also seems clear that human moral behaviour goes further.

How far can morality be explained in terms of evolution? Clearly, in some sense, morality has evolved. However, there has long been a vigorous debate among scientists about whether morality cuts with the grain of evolution, or goes against it.

On the one hand, morality can be seen as having arisen from evolution, and been favoured through natural selection. On the other hand, it can be seen as running counter to the priorities of natural selection, and an indicator that selection pressures are not as dominant in human beings as they have been in other species, given our greater mastery of the environment. In the 19th century Herbert Spencer thought the former, Thomas Henry Huxley the latter. In recent sociobiology, Edward O Wilson is inclined to see morality as arising from evolution, but Richard Dawkins, in *The Selfish Gene*, quietly abandons that position.

Those who espouse evolutionary ethics often assume that there could not be any external source of moral norms. However, that does not necessarily follow from the evolutionary approach. It is perfectly possible to hold some kind of "natural law" approach to ethics, whether or not that is based on the will of God, but also to claim that the human capacity to apprehend moral norms has evolved. Sometimes, evolutionary ethics takes the additional step of suggesting, not just that ethics is explicable in terms of evolution, but that it should be reformulated in terms of evolutionary principles. That is moving from an explanatory to a prescriptive use of evolution, and is clearly not a necessary consequence of accepting that human moral capacities have evolved.

Evolution has given rise to a general debate about the degree of similarity between human beings and other higher animals. Those who want to exaggerate the extent to which human beings are "nothing more" than the products of evolution tend also to want to exaggerate the similarity between humans and other animals. Clearly, there are both similarities and differences, and it is important to keep them in balance, rather than to exaggerate one or the other. Secular thinkers tend to exaggerate the differences, while religious thinkers exaggerate the similarities – though there is no reason that should be the case. There is no reason why humanists should not recognise what may be distinctively human. Equally, it is certainly part of the Old Testament tradition to recognise the similarities between humans and the "beasts" (e.g. Ecclesiastes 3), while also recognising the special importance of human beings in the purposes of God. For Christians, the latter point is particularly linked to the recognition of God's incarnation in human form.

Most of the higher attributes of human beings can be found in less developed form in other animals. Language has been one of the most hotly debated cases. Though some primates have a capacity for using signs that borders on language, the majority scientific view is currently that it is an exaggeration to say that they actually have a capacity for language. The same is probably true of the capacity to form relationships. There is clearly a capacity for bonding in higher animals, but the human capacity for relationship seems to be of a different order. Many animals are clearly conscious of their environment, but human beings have a higher-order consciousness that involves *knowing what they know*. That higher consciousness is now widely recognised to be the key feature of what is distinctively human.

Salvation History

Darwin seemed to have set out a view of the world in which there was no place for Christ and, in that sense, Darwinism seemed to be a threat to Christianity. However, we have now developed an understanding of the place of Christ in evolution, and that has really enriched our understanding of his significance. The integration of evolutionary theory and Christian salvation history has been a particular feature of British theology, and in recent decades Arthur Peacocke (e.g. 1993) has been its most important exponent. In Catholic thought, Karl Rahner (1978) has also made a significant contribution.

The first question is to know how to place the Fall in relation to evolution. Most theologians assume that the story of Genesis 3 was not intended to be taken literally, and so should be interpreted as making general points about human nature. While that is a perfectly defensible approach, it is also possible to see Genesis 3 as describing, in narrative form, the long-term evolutionary trend that must have resulted in the development of human moral consciousness, or the "knowledge of good and evil".

Such a reading of Genesis 3 will see the development of moral consciousness as being in some sense a "fall upwards". Though it represents an additional human capacity, it makes it possible for humans to do evil with a new kind of deliberateness, and that is close to the heart of what is meant by the concept of "sin". Some may feel that this evolutionary reading of Genesis 3 does not giving a sufficiently central place to sin. However, that perhaps arises from the sin-laden reading of the text that has been current in Western Christianity for many centuries, but is not found in Eastern Christianity, or in Judaism. As James Barr (1992) has pointed out, the text itself says almost nothing about sin.

Darwin has drawn attention to the struggle for survival in nature. Tennyson spoke of nature "red in tooth and claw". Before Darwin, many people had clung to a romantic view of nature in which they had averted their eyes from the struggle for survival, and the evil and suffering that goes with that (though there were exceptions such as Malthus who had influenced Darwin). Darwin made people aware of how much struggle takes place in the natural world. At first glance, that seemed to be a threat to Christianity. However it was only a threat to the ill-considered Christianity that wanted to pretend that everything in nature was good. Exactly how evil has entered into creation is one of the great mysteries. However, it is an evident fact that good and evil are intertwined in creation.

The next question is how to place Christ in relation to evolutionary theory. The general tendency has been to see Christ as marking the culmination of natural evolution. Evolution has led to creatures in whom it was possible for God to become incarnate. The incarnation can thus be seen as, in one sense, the goal of evolution. Equally, evolution makes it possible to give an account of the "finality" of Christ. This all seems satisfactory as far as it goes, but more is needed for an adequate Christology (Polkinghorne 1996). Another step can be taken by suggesting that Christ reveals God's purpose in evolution and, through that revelation, affects the future course of evolution. That is also unproblematic, but may still not be quite enough. It is difficult to suggest convincingly how Christ has had an impact on the course of evolution, though some attempts have been made to do that. Gerd Theissen provides one approach, suggesting that Christ marks a new "spiritual mutation", a novel form of humanity that unlocks possibilities for the future (Theissen 1984).

One weakness of most evolutionary Christologies is that they tend to be too indebted to background assumptions about human progress. Evolutionary theology had its origin in an age in which assumptions about human progress were very strong, and it was tempting to see both evolution and salvation history as providing narratives about progress. The task then seemed to be simply mapping one on to the other. The problem with this approach, as John Polkinghorne (1996) among others has pointed out, is that it does not give an adequate account of the significance of Christ, or of what difference he makes to salvation. If things were getting better in the lead-up to Christ, and continue to get better in the time after Christ, it is not easy to argue that Christ made any fundamental difference.

I would suggest this is a complaint about how evolutionary theology has usually been approached, rather than a criticism of the whole project. I would be inclined to see Christ as marking a turning point in the development of human consciousness as, for example, Rudolf Steiner and those influenced by him, such as Owen Barfield (1977) suggested. In the centuries leading up to Christ there seems to have been a gradual loss of the animistic consciousness that effortlessly saw God as speaking to humanity through the natural world. To replace that fading animistic consciousness, Christ gave the Spirit to dwell within humanity. This marked a fundamental change in the sense of where the spirit was located – no longer beyond nature, but within humanity. Christ can thus be seen as a turning point in the evolution of human consciousness. This approach illustrates how it is possible to formulate an evolutionary theology that allows Christ to be seen as making a fundamental difference to the course of events, rather than just a staging point in an "onwards and upwards" progress.

Evolution of Religion

Darwin had relatively little to say about religion, though he mentions it briefly in the last chapter of *The Descent of Man* (1871), and places it in the evolutionary context. It is only recently that the evolution of religion has become an active field of evolutionary theorising, and the full implications for our view of the nature of religion are not yet apparent. Many of those working on the evolution of religion approach their task with atheistic assumptions, and rather too readily assume that the evolutionary approach supports that. However, it does not necessarily do so, and it is possible for an evolutionary approach to be welcomed by those sympathetic to religion as providing a helpful, critical perspective.

There is no problem about the basic claim that religious capacities have evolved through evolution. They seem closely linked to the capacities for language, self-consciousness and relationality that are distinctively human. Issues arise about whether or not religion has been good for survival, but even those who dispute the claims of religion can see that, through promoting good social behaviour, it may have facilitated survival. However, it would be quite another matter, and an illegitimate inference, to assume that religion is "nothing but" the product of evolution. It would also be wrong to think that the evolution of the human capacity for religion bears one way or the other on the truth of belief in God. It is entirely consistent to think that God intended that evolution should lead to creatures such as ourselves who have a capacity to believe in him and to worship him.

The main facts about the evolution of religion are clear, and I don't think there is any dispute about them. As Steven Mithen (1996 and this volume) says, remarkable developments took place for which there is clear archaeological evidence. 200,000 years ago, *Homo sapiens* appeared. First, they started burying their dead. Then, 60,000 years ago, they made boats and a much wider variety of tools. Gradually, they constructed dwellings, painted walls, made carvings, planted crops etc. Religion and art developed. Everyone agrees that something fundamental changed at the Middle/Upper Palaeolithic transition, though this is difficult to date exactly because it occurred at slightly different times in different places.

I am very much influenced by Mithen's idea that what made all this possible was a kind of fluid intelligence that did not respect boundaries between domains (animate and inanimate). Similar ways of thinking are applied to both the animate and the inanimate domains. My problem with that is what I believe to be an unsubstantiated mythology about what came before this. We know very little about that, as Mithen admits. It is frequently assumed that before cross-domain thinking, there was a prior period in which people could tell one domain from another. Similarly, it is assumed that before "counter-intuitive" domain violation, there was a prior period in which emerging humanity stayed within "intuitive" assumptions. I want to question those assumptions about a prior period of domain-specific thinking, and about what is "intuitive" and what is "counter-intuitive". I suggest that they may be driven by modern ideas about what is normative, and represent a projection back into evolutionary history of the assumptions of contemporary atheism.

The prevailing story about how counter-intuitive thinking arose strikes me as like a myth about a "Fall". It is like the myth of the Garden of Eden, but a myth driven by the assumptions of modern, materialist atheism, rather than those of primitive religion. The Garden of Eden myth imagines a time when God and humanity were on easy and intimate terms, when life was easy, and when toil, tears and mortality had not yet begun. I very much doubt whether that corresponds to the reality of the evolution of humanity. For evolutionary atheism, the equivalent of the Garden of Eden is when counter-intuitive thinking had not yet begun. They look back to a time when humanity had an unwavering recognition of the purely material status of the natural world, and the pernicious assumption that it came from a living God had not yet arisen. I suggest that the myth of a loss of domain-specific rationality may bear little more relation to what actually took place in evolution than the Eden myth of Genesis 3 [2].

There seems to be a temptation in mythological thinking to trace the evidence back as far it will go and then to assume, when the evidence peters out, that before that things were completely different. Life for emerging humanity seems to have been nasty, brutish and short for as far back as we can trace it. Strange, then, to assume, as the Garden of Eden myth does, that there was an earlier period in which everything was lovely and easy. Equally, as far back as the evidence will take us, we can find cross-domain thinking. The further back we go, the stronger the evidence for that seems to be. Strange then to assume that there is a point further back, beyond the evidence, where things were completely reversed.

I was alerted to the oddness of these assumptions by an old debate about the origin of metaphor. It is often assumed that language was literal before it was metaphorical. Max Muller was one of the original advocates of that position, and it still seems to be widely accepted. However, as Owen Barfield (1928) pointed out, it sits very uneasily with the evidence from historical etymology. As far back as we can trace, the meanings of words are increasingly metaphorical. It is very strange then to assume that there is a prior period, for which there is no evidence at all, when language was purely literal, and metaphor had not yet begun. In the 1950s and 60s, the social psychologist Solomon Asch (1958) provided additional support for the metaphorical origins of language by demonstrating that, in historically independent languages, you get the same "dual-aspect" terms, as he called them, linking aspects of the natural world and aspects of personal experience.

The idea of a purely natural world, independent of agency is, I suggest, a modernist idea, projected back into evolutionary origins. There is another viewpoint from which it is a very odd idea. Since Kant, philosophers have accepted that the reality we know is a constructed one. More recently, cognitive psychologists and social constructionists have, in their different ways, elaborated that assumption. Despite the widespread acceptance of this recognition, it seems to co-exist with the assumption that the material world, as *we* construct it now – in our modern agency-free way – is exactly how it was really encountered, as humanity was evolving. It is further assumed that emerging humanity must have constructed the world exactly as we do, before the "fall" into domain violations and breaches of what we now regard as "intuitive" assumptions.

---

[2] Indeed, their shared theme, of the loss of something akin to innocence, suggests that they may have more in common with each other than with reality! (Ed)

There is perhaps one aspect of the Garden of Eden myth that corresponds to evolutionary reality. That is the strand concerned with the development of conceptual distinctions. The story hangs around our ancestors' acquiring the distinction between good and evil. There is also, implicitly, the development of a parallel distinction between God and humanity. It seems to me highly likely that, in this regard, the myth of the Garden of Eden is describing an evolutionary development that actually took place. Another of the key conceptual distinctions that developed, I suggest, is the distinction between domains. The acquisition of such conceptual distinctions was, as implicitly acknowledged above, a kind of fall "upwards", with huge practical advantages, though perhaps accompanied by some sense of loss in the form of a sense of disconnectedness and alienation.

It is one of the interesting features of the current evolutionary/cognitive science approach to religion that it recognizes the naturalness of religious thinking. Pascal Boyer called his first book *The Naturalness of Religious Ideas* (Boyer 1994). It seems to me to sit strangely with the idea that religious ideas are natural to also say that they are counter-intuitive, and I am not convinced that it is a circle that he satisfactorily squares. The assumption that religious ideas are natural is one that I share with the prevailing orthodoxy. It would not be appropriate to build any religious argument on that basis. However, at very least, it is not inconsistent with what might be expected from the religious assumption that our minds and brains exist in the context of a wider theistic reality. Though the evolutionary approach to religion is currently intertwined with atheism, I suggest that evolutionary data are compatible with a much more positive view of religion than is normally imagined. Though these are early days in the evolutionary approach to religion, it is an approach that may yet enrich our understanding of the nature of religion in ways that are constructive and helpful.

Conclusion

Evolutionary biology has been at the forefront of the interchange between science and religion for a century and a half. This has focused around a variety of different concerns. One has been whether the view of origins presented in the book of Genesis is compatible with evolutionary biology. Another is whether there is evidence of the kind of directional change in evolution that is required if it is to be interpreted as an unfolding of God's providence. Yet another has been the extent to which human beings can be understood and explained in terms of their evolutionary origins, and whether the evolutionary approach is compatible with the Christian view, both of human nature in general, and of morality and religion in particular. Further, there is the question of how far it is possible or desirable to reframe Christian doctrine in the light of evolutionary biology. For the most part, the dialogue between theology and science has focused rather generally on creation and providence, and it is only in dialogue with evolutionary biology that there has been any attempt to rethink the salvific role of Christ. It has been argued here that, in each case, it is not just that there is no conflict between evolution and Christian theology, but that the rethinking of Christian theology to which Darwinism has led has been entirely beneficial.

The focus of debate is now shifting to views of religion, and it is often assumed that an evolutionary approach to religion is most readily compatible with atheist assumptions. However, there is reason to think that those assumptions are, in some ways, distorting our understanding of the evolution of religion. When those distortions are cleared away, a view of the evolution of religion may emerge that is positive about religion. An instructive parallel is that Freud took a negative view of religion, but subsequent psychoanalytic theorists have taken a much more positive view of it. However, that task of rethinking the evolutionary approach to religion is a larger one than can be carried through in a single section of this short chapter.

Bibilography

Asch, S.E. 1958. The Metaphor: a Psychological Inquiry, in R. Taguiri & L. Petrullo (eds) *Person, Perception and Interpersonal Behaviour*. Stanford: University Press.

Ayala, F.J. 1988. Can Progress be Defined as a Biological Concept?, in M.H. Nitecki (ed) *Evolutionary Progress*. Chicago: University Press.

Barfield, O. 1928. *Poetic Diction: A Study in Meaning*. London: Faber & Gwyer.

—. 1977. Philosophy and the Incarnation, in O. Barfield, *The Rediscovery of Meaning and Other Essays*. Middletown, Conn.: Wesleyan University Press.

Barlow, C. 1994. *Evolution Extended: Biological Debates on the Meaning of Life*. London: MIT Press.

Barr, J. 1992. *The Garden of Eden and the Hope of Immortality: The Read-Tuckwell lectures for 1990*. London: SCM.

Boyer, P. 1994. *The Naturalness of Religious Ideas: A Cognitive Theory of Religion*. London: University of California Press.

Brooke, J. H. 1991. *Science and Religion: Some Historical Perspectives.* Cambridge: University Press.

Conway Morris, S. 1988. *The Crucible of Creation.* Oxford: University Press.

Darwin, C. 1871 *The Descent of Man.* London: John Murray

Kauffman, S. 1995. *At Home in the Universe.* London: Viking.

Mithen, S. 1996. *The Prehistory of the Mind: A Search for the Origins of Art, Science and Religion.* London: Thames & Hudson.

Monod, J. 1972. *Chance and Necessity.* New York: Vintage Books.

Moore, J. R. 1979. *The post-Darwinian Controversies: A Study of the Protestant Struggle to come to terms with Darwin in Great Britain and America, 1870-1900.* Cambridge: University Press.

Paley, W. 1802. *Natural Theology.* London: Rivington.

Peacocke, A. 1993 *Theology for a Scientific Age (edn 2)* Oxford: Blackwell

—. 2001. *Paths from Science towards God.* Oxford: Oneworld.

Polkinghorne, J.C. 1996. *Scientists as Theologians: a comparison of the writings of Ian Barbour, Arthur Peacocke and John Polkinghorne.* London: SPCK.

Rahner, K. 1978. *Foundations of Christian Faith: An Introduction to the Idea of Christianity.* London: Darton, Longman & Todd.

Ruse, M. 1995. *Evolutionary Naturalism.* New York: Routledge.

Theissen, G. 1984. *Biblical Faith: an Evolutionary Approach.* London: SCM.

Thompson, P. 1995. *Issues in Evolutionary Ethics.* Albany: State University of New York Press.

Watts, F. 2002. *Theology and Psychology.* Aldershot: Ashgate.

CHAPTER FIVE

TIMEO DARWINOS ET DONA FERENTES[1]: A RESPONSE TO FRASER WATTS

ANTHONY FREEMAN

Revd Anthony Freeman read chemistry and then theology at Oxford, befoire being ordained into the Church of England in 1972. However, he was dismissed from his parish in 1993, for the views he put forward in his book, "God in Us: A Case for Christian Humanism", which represents God as the sum of our individual ideals and values, not as an invisible person somewhere "out there". He remains a priest, but makes his living as a lecturer, writer and managing editor of the Journal of Consciousness Studies. He has also written "Gospel Treasure" (1999) and "Consciousness: A Guide to the Debates" (2003). A 2nd edn of "God in Us" appeared in 2001.

In this response he argues that Fraser Watts takes in general too generous a view of Darwin's legacy for Christian belief, then explores further the implications of Darwinian thought for our understandings of Christ on the one hand, and of evil on the other.

Plan of response

I shall reflect on Fraser Watts' paper under the same heads that he uses, considering first the four areas in which he claims that Darwinian evolution has proved a good friend of Christianity, and then turning briefly to his discussion of Darwin's more general influence on our approach to religion.

---

[1] "Beware Darwinists bearing gifts" – with apologies to Virgil.

Bible

From the sixteenth to eighteenth centuries, the new learning of the Renaissance and the Enlightenment had resulted in a shift from the fourfold exposition of scripture – literal, allegorical, tropological (moral), and anagogical (mystical) – favoured in the Middle Ages. By the mid-nineteenth century these complex and often arbitrary interpretations had given way to a more historically-informed style of exegesis. Some exegetes in the first half of the nineteenth century were already *de facto* literalists, simply by dint of having dropped the other three elements of the earlier system; but Fraser is surely right that the dogmatic insistence on an exclusively literal interpretation of Genesis 1 & 2, associated today with Biblical fundamentalism, came about as a response to Darwin's evolutionary theory.

What surprises me is the suggestion he appears to make that this is among the things for which Darwin should be thanked. How can this literalist development be seen as anything other than a disaster for Christianity in general and Biblical studies in particular? Of course, he does not refer to this starkly, in isolation: he begins with Aubrey Moore, and later quotes Charles Kingsley, in arguing for the indwelling of the created order by an ever-present divinity. On these and other grounds he concludes his section on the Bible by saying "it is arguable" that Darwin facilitated the growth in a more sophisticated way of understanding the Bible. Yet even on his own brief account of the process, the opposite is the ultimate case. Had Darwinism not provoked the literalist backlash against the historico-critical study of the Old Testament, the higher criticism would have been accepted and we should not be suffering the scourge of Biblical fundamentalism today.

Creation, Progress and Providence

Fraser describes as "heretical" the deist view that "God had created things once for all at the beginning", a doctrine widely held by English divines and senior Churchmen in the eighteenth century. Then he credits Darwin with having "helped people to understand that creation is an ongoing project for God", so weaning them off deism. It seems to me that rather than set deism against Darwinism, Fraser would do better to combine them, presenting Darwinian evolution as the mechanism whereby the deist's creation is managed. If we understand evolution in a strict Darwinian sense – rather than as a vague term for development – then God cannot be in detailed control of the ongoing creative process, dependent as it is on chance mutation and natural selection. If God is the creator of Darwin's evolving world, then it is precisely in the sense that he set the starting conditions to bring it about – and that is the essence of deist doctrine of creation. Viewed in this way, whether or not "directional change" should turn out to be a reality, the course of evolution supports Kingsley's "loftier thought" concerning God without requiring belief in day-to-day divine intervention.

It is true that some recent Christian thinkers have proposed ways in which – they contend – a creator God could continue to direct, or fine-tune, the course of evolution, without intervening directly in individual developments. Particularly often cited, in this connection, is the late Arthur Peacocke (1979, 2004), whose role in establishing the Science and Religion Forum is widely appreciated. Others explicitly arguing in such terms include Nancey Murphy (1995), John Haught (2000) and John Polkinghorne (2000). But none of the mechanisms proposed seems to me convincing – and, in any case, Fraser doesn't refer in this connection to any of these writers, while those he does mention, Conway-Morris and Kaufmann, appear even vaguer. So I remain unpersuaded that he in fact progresses the argument beyond deism.

There is however a further hurdle in the path of a reconciliation between Darwinian evolution and the Christian doctrine of creation. The inherent viciousness of nature – red in tooth and claw – is today the chief stumbling block to belief in a good Creator. Traditional Christianity had shifted the blame for so-called natural evil from the Creator to human sinfulness, but Darwin denies us this escape route. Not only does he present the fight for survival positively, as fundamental to the success of an evolving creation, he also removes humankind from the pole position that allowed theologians to argue that human frailty was the cause of this characteristic harshness of the natural world. These matters are treated further in the next two sections.

Human Nature

When Disraeli asked rhetorically, "Is man an ape or an angel?", and declared himself "on the side of the angels", he was nowhere close to the Biblical doctrine of human nature. However, the Darwinian view that he repudiated, and which Fraser Watts seems to espouse, presents a dethroned humanity that is equally inconsistent with traditional Christianity. Fraser may be right that the assumptions about human dignity that were challenged by Darwinism were never *specifically* Christian, but Christian they undoubtedly were and are. Humans were put on their pedestal by the

Hebrew Bible and kept there by the Christian doctrines of Fall and Redemption. There was nothing comparable in other writings of antiquity.

Darwin certainly taught us to see the human race in a quite new perspective – and one which may yet prove to be compatible with the high doctrine of mankind found in the Biblical and Christian traditions – but to claim as Fraser does that this apparent downgrading of our species makes Darwin a friend to Christianity seems to me perverse.

Salvation History

An approach to Darwinian evolution that sees in it a preparation for the Incarnation of God in Christ, especially with regard to the evolution of consciousness, is one that I am happy to share with Fraser. But I am even more of a heretic that the deists, and he should ask himself carefully whether he really wants me for company. My own evolutionary interpretation of Jesus – evolving as both Man and God – treats God-consciousness and indeed Godself as emergent states or properties (Freeman 2001), and Fraser has confirmed in public discussion that he would not accept that line of argument. But, as he acknowledges in his paper, even more modest proposals in this direction require that Christ's humanity has evolved to achieve that special place in creation traditionally described as being "in the image of God". This is what makes the incarnation possible. This brings us back to the question of how far and in what way God can be said to be in control of the evolutionary process. To stand in any continuity with traditional Christianity, a Darwinian interpretation of events must see God as in some sense the author of the incarnation. It cannot result purely from chance.

A second issue is the uniqueness of Christ. Here Fraser might do well to look to past figures such as Abelard and Schleiermacher, in whose theologies the Redeemer does not cut against the grain of nature in a unique way, but is rather the logical and natural fulfilment of all that is best in it. Such a Jesus does not need to be radically different from others, rather he becomes the iconic model of perfect humanity and because of this (not in spite of it) of God as well. As Karl Rahner (1961 p.184) put it:

> Only someone who forgets that the essence of man is to be unbounded . . . can suppose that it is impossible for there to be a man, who, precisely by being man in the fullest sense (which we never attain), is God's existence in the world.

Fraser is likely to find such an approach over-optimistic, because for him "it is an evident fact that good and evil are intertwined in creation".

Here I disagree. Good and evil are certainly intertwined in human moral consciousness, but that does not mean they exist objectively in nature. Floods and earthquakes occur, but only become disasters when they are perceived as such by sentient beings whose lives they affect in negative ways, and they only become moral evils when designated as such by human beings. A cat kills and eats a bird, which is bad for the bird but good for the cat, and good also for the unsuspecting worm the bird was in turn about to catch and eat. The event may be regarded as good and/or bad, depending on one's point of view, but in itself it is morally neutral. So there is no good creation in need of redemption from evil; the same evolutionary development which brings an appreciation of morality brings also the resolution of its perceived problems as conscience and consciousness evolve further.

Evolution of Religion

Moving to the final section of Fraser Watts' paper, on the evolution of religion, I have really only one point to make. If we accept – as Fraser seems to do – that both the modern agency-free view of the universe and the earlier divine-guidance view are the products of naturally evolved human brains and the thoughts to which they give rise, then important questions arise. For instance:
- Is there an unchanging reality independent of these ideas?
- If so, is there any view-from-nowhere to access it?
- If not, are there other criteria for preferring one account to the other?

Or taking up his point about the development of conceptual distinctions:
- Can good & evil or human & divine exist apart from our consciousness of them?
- Are the distinctions between good/evil or divine/human essential to these concepts?

Finally, I do agree with Fraser that evolutionary atheism is akin to a religious belief, with its own myths and dogmas. I commend philosopher Bill Robinson's (2007) useful distinction between *Biological* Evolutionary Theory (which says the physical constitution of complex organisms is shaped by natural selection operating on variability) and what he labels *Ideological* Evolutionary Theory (which claims that evolutionary considerations explain everything that exists now but did not exist a billion years ago). *Pace* Richard Dawkins, the first theory does not entail the second.

References

Freeman, A. 2001, God as an emergent property, *Journal of Consciousness Studies*, **8** (9-10), 147-159.

Haught, J.F. 2000. *God after Darwin: A Theology of Evolution.* Boulder, CO: Westview Press.

Murphy, N. 1995. Divine Action in the Natural Order: Buridan's Ass and Schrödimger's Cat in *Chaos and Complexity: Scientific Perspectives on Divine Action.* Vatican City / Berkeley, CA: Vatican Observatory / Centre for Theology and the Natural Sciences, 325-357.

Peacocke, A.R. 1979. *Creation and the World of Science,* Oxford: University Press

—. 2004. *Evolution, the Disguised Friend of Faith?* Philadelphia: Templeton Foundation Press

Polkinghorne, J. 2000. *Faith, Science and Understanding.* London: SPCK

Rahner, K. 1963. *Theological Investigations*, **1**, Baltimore, Md.: Helicon Press.

Robinson, W.S. 2007, Evolution and epiphenomenalism, *Journal of Consciousness Studies*, 14 (11), 27-42.

CHAPTER SIX

WHAT CAN EVOLVED MINDS KNOW OF GOD? RECONSIDERING THEOLOGY IN THE LIGHT OF EVOLUTIONARY EPISTEMOLOGY

NEIL SPURWAY

Neil Spurway studied physiology to PhD level in Cambridge. His professional life was spent in the University of Glasgow, retiring as Professor of Exercise Physiology in 2001 and as Chair of the British Association of Sport and Exercise Sciences a year later. Interested from schooldays in the Science/Religion dialogue, he has also chaired Glasgow's Gifford Lectureships Committee, and edited ESSSAT News, membership journal of the European Association for the Study of Science & Theology. He is currently a Vice-President of that body, and Chair of the Science & Religion Forum. His most recent books are "Genetics and molecular biology of muscle adaptation" (with H. Wackerhage, 2006) and "Creation and the Abrahamic faiths" (ed., 2008), the predecessor of this volume.

Here he argues, from the standpoint that human minds are the products of evolutionary selection on the surface of this earth, that they should never claim certainty, or anything approaching it, in their beliefs about God, or any being or state of existence outside space and time.

Biology as a starting point

Socrates, as retailed to us by Plato, proclaimed that: "The unexamined life is not worth living". Anyone reading this book obviously agrees, and has probably come up with 10,000 questions during his or her lifetime. Answers, however, are more elusive; where should we even begin, amid the flux of uncertainties?

Descartes' attempt to identify an unchallengeable base-point *(Cogito, ergo sum)* is now unfashionable. First year philosophy students are taught to pick holes in it. Yet there is no other agreed foundation – no "first philosophy" to which all rational beings may be expected to assent. Indeed many recent thinkers have striven to be "nonfoundationalist", and treat every problem in its own historical, societal or disciplinary terms. That way, however, cannot lead to consistency even within one person's mind; consensus extending over large groups will be quite impossible on any question other than the impossibility of consensus.

If we are to do better than this we must find what we should perhaps avoid calling a "foundation", because of its historical connotations, but can at least accept as a common starting point. Having myself worked in a biological science I am happy to notice that an increasing number of thinkers, in a range of disciplines, are adopting Darwinism as such a starting point. Arthur Peacocke (1979), Edward Farley (1990), Philip Hefner (1993), John Haught (2000), Gordon Kaufman (2004) and Wentzel van Huyssteen (1999, 2006) are among theologians of whom this is true. My focus here is on van Huyssteen who has stressed at many points that the implications of evolution for our understanding of knowledge itself – as considered in the discipline of Evolutionary Epistemology (EE) – must be most seriously assessed by theologians. Three quotations from his Gifford Lectures (van Huyssteen, 2006; page nos in brackets) document this position:

> ... theology has traditionally virtually ignored the question of the evolution of human cognition (311)

Yet:

> ... it would be a serious mistake to think that ... one could conceive of an epistemology independently of biology." (283)

So:

> ... for theologians the following should be true: if we take the theory of evolution seriously, we should take evolutionary epistemology seriously. (85)

I welcome and wholeheartedly endorse van Huyssteen's challenge, but argue in this paper that he just misses the most critical implication of EE. Van Huyssteen, and to lesser extents predecessors such as Pascal Boyer (2001), David Sloan Wilson (2002) and Justin Barrett (2004), have recruited EE in considering how God-concepts arose. However, I go further, and ask not only what EE can suggest about how such concepts came into being, but *what guidance it can give about whether they may be trusted.*

Thus I seek to formulate, not my theology itself, but my appraisal of *what kinds of thing theology can properly say*, in the light of biological, Darwinian thinking.

Evolutionary Epistemology

The basic contention of EE is that not only our bodies but our brains, and not only our brains as material entities but the kinds of concept they can form, are all entirely products of natural selection. Such thinking originated with Charles Darwin himself (1871 and *Notebooks,* surveyed by Gruber, 1974), William James (1890), Georg Simmel (1895), Herbert Spencer (1897) and a number of other biologist-philosophers around the turn of the 19th into the 20th centuries (further refs in Campbell 1974). It was brought into prominence again in two rather different ways by Konrad Lorenz (1941, 1977) and Karl Popper (1972), respectively. Among those who have further developed the ideas involved have been Donald Campbell (1974), Gerhard Vollmer and other contributors to Wuketits (1984), Peter Munz (1985, 1993) and Henry Plotkin (1994).

The most obvious respect in which this evolutionary claim is true is that of our sense organs (Barlow & Mollon 1982). As these are the routes by which we acquire all our knowledge, they are critically important. Consider vision. Our eyes can detect only a narrow range of the electromagnetic spectrum, yet it is a range to which most material objects which we might wish either to avoid or grasp are opaque, so by utilizing this waveband we can see them. Our brightness-sensitivity self-adjusts over a range of about 10^{12}, from the strongest sunlight to a very dark cave; and when maximally dark-adapted our sensitivity reaches the ultimate physical limit, one photon being sufficient to excite a rod. Yet there is a further subtlety which underlines the evolutionary honing: a single rod, in that state of peak sensitivity, could too easily discharge by thermal accident, or under the impact of a blood cell in an adjacent capillary, so in fact we only notice a momentary flash if several neighbouring rods fire almost simultaneously – the number and collecting-time being within the optimum range demonstrated by information theory (Hecht, Schlaer & Pirrenne 1942; Brindley 1960).

For hearing there is no similarly absolute physical limit, but we can hear sounds sufficient to vibrate our ear-drums only $1/10^{th}$ the diameter of a hydrogen atom – yet only in a frequency-range safely above those of body noises, which otherwise would deafen us. On the other hand, predatory birds which hunt from many tens of meters above the ground have higher visual acuity, many animals can hear higher frequencies, our touch and smell capacities fall far short of those other species to which one or other of these sensory modes matters more, and we have no capacity to detect electrical signals as some fish do, the polarization of light like some insects, and so on. As many as 80 years ago, von Uexküll (1928) designated the contrasting sensory worlds of the various species as their *Umwelten* (translated by Vollmer as "ambients"). Just as significantly, there is a huge range of incident energies, all potential sources of information, which no species can access – for example, most of the electromagnetic spectrum. Usually in these cases the psycho-physicist can see that the bodily dimensions of animals, or the materials of which they can be made, preclude the wider range. There also remains the very considerable possibility that there are potential information-sources of whose existence we have no idea. Nonetheless, even the inter-species comparisons which we can make show that each one could, in principle, have wider awareness than it does. What determines the capacities each in fact possesses? – The ratio of evolutionary benefit to developmental cost.

> We have developed "organs" only for those aspects of reality of which, in the interests of survival, it was imperative for our species to take account
> (Lorenz 1977)

Similar arguments apply to brains. Much of the detail is highly technical, but some points are easily described. For instance, in human beings, the areas of the sensory cortex devoted to thumb and finger tips, and to lips, are much larger than those to limbs and trunk – the relevant factor being the importance of tactile discrimination in the respective body parts, not their own surface area. And a pig's snout has a brain-region several times larger than our fine-touch areas. Not that it should be assumed that the mechanisms determining innervation density are entirely genetic. A foetus a few weeks before term has many more nerve synapses (sensory and motor) than it will have a year or two later: extensive perinatal elimination of less useful synapses occurs in parallel with reinforcement of others. This process has been widely observed in the peripheral nervous system, and undoubtedly occurs in spinal cord and brain too (Edelman 1989). Such "Neural Darwinism" is still a process of natural selection, yet it is not only subtly different in each individual, it is also many orders of magnitude faster than genomic evolution.

> Interactions with the environment contribute to the formation of more and more complex neuronal organization .... Each generation renews this

selective shaping of the brain by the environment. It is accomplished very rapidly compared with the geological time scale of the genome's evolution. [This] epigenesis by selective stabilization saves time. The Darwinism of synapses replaces the Darwinism of genes. (Changeux 1986)

However, whichever the mechanism of selection, we may be sure that both the information captured and the data-processing capacity applied to it are biologically determined by their cost/benefit ratio.

Adaptation

The fundamental driving force behind the mechanisms just considered was put "crudely but graphically" by Simpson (1963):

> The monkey who did not have a realistic perception of the tree branch he jumped for was soon a dead monkey – and therefore did not become one of our ancestors.

Strictly speaking, more recent studies have shown that monkeys, in the modern sense, were not directly on our evolutionary line – rather, they and we share a common precursor, and it is to that precursor that Simpson's contention applies. But that does not affect his argument. And what is true of spatial and other physical judgements must, from a Darwinian standpoint, be true of biological and social judgements also. The consequences of being wrong would in many cases have been fatal, and in all others they would have been failure to thrive in the competitive world. But we *have* survived, and emerged from a network of ancestors who survived where others failed, so we may conclude that our objective concepts have been highly trustworthy and our social dispositions, over the course of history, on balance valuable.

Of course, we are not born with detailed knowledge of our individual environments, let alone anticipation of the uncountable number of events which will occur in a given life. What is conveyed in our DNA, and more finely tuned by synaptic competition, is responsiveness to particular forms of regularity in our environments – at first in evolutionary time this will have been entirely the physical environment, in the later stages it must increasingly have been the social one. Yet, where the implications for survival were definite, the mechanism will have been essentially the same – relative or absolute success or unsuccess in the competition to reproduce. An example of the first kind is that nervous systems are not predisposed to judge the leap to a particular branch, but to make spatial judgements generally. An instance of the second kind is that the brains of human infants are not adapted to the learning of English in one child and Chinese in another, but in each to the learning of language in the broad. However, the extraordinary speed with which we do learn the language(s) heard all round us in our early years must give the strongest indication to any doubter that the propensity for such learning is, indeed, inborn.

Looking back to the previous quotation from Lorenz, it seems inescapable that not only our physical but our mental "organs" – our capacities to undertake particular forms of mental process and formulate particular kinds of concept – have been selected by their contributions to our ancestors' survival and reproduction. Thus, in instances where survival is directly affected, not only perception but also *con*ception must be pretty accurate.

- If we were wrong in our spatial judgements, swinging between branches or leaping over chasms, we would either fall, probably fatally, or be likely to concuss ourselves against the far side.
- If wrong in our concepts concerning other animals we would be prone on the one hand to try and cook logs or even rocks, on the other to try and cuddle tigers or mate with gorillas.
- Further physical concepts would be just as critically tested in our avoidance of very hot or very cold things, and of putrefying, physically unstable or otherwise hazardous objects or environments.
- Subtler, social interactions, from an inhibition against biting or kicking everyone we met to dispositions toward cooperation, altruism, sensitivity to others' feelings, and even love, could easily be in the longer run just as critical.

If it were possible to set up a calculus of credibility, the social concepts which have guided us would surely all score over 50% and the physical ones close to 100%. And the origin of them all? – Biological selection, survival of the fittest. No other mechanism is evident or needed. So:

> Evolutionary Epistemology claims that … human cognition arose according to natural laws. No miracle was necessary, no divine intervention, no offence against the laws of nature. (Vollmer, 1984)

Plato and Kant

Readers with any knowledge of the history of philosophy will by this point have perceived that EE represents a distinctive position in relation to the debate, which has extended from Plato to the modern era, about the existence or otherwise of innate ideas. Empiricists, of whom probably the most extreme was Locke, have always denied that we have such ideas – the mind of a newborn baby is, to them, "*tabula rasa*", a clean slate. The opposite school, broadly classed as Idealists, have maintained that we perceive the world in terms of ideal concepts with which we are born. The earliest fully recorded version of this stance is that of Plato. But Darwin himself commented in his Notebooks:

> Plato says in Phaedo, that our imaginary ideas arise from the pre-existence of the soul, are not derivable from experience. Read monkeys for pre-existence!

Before this, he had already said,

> He who understands baboon would do more towards metaphysics than Locke. (Both quoted in Gruber, 1974)

An account of innate ideas more acceptable to the relatively modern mind was that of Kant. For him concepts such as of space and time (concepts fundamental to Newtonian physics) are given to us "*a priori*": that is to say, they are not products of our personal experience, but instead are the basic mental "givens" in terms of which we interpret all experience. But this has the unhappy consequence that, since we can never assess the world without these *a priori* concepts, we can never know if they are right. Nor can Kant's approach explain how such concepts arise. William James, at the turn of the 20th C, and more recently and consistently Konrad Lorenz, argued that they are not mysterious, but products of Natural Selection – and consequently trustworthy: though they are almost certainly not perfectly correct (and if they were we could never know it), we do know that they are sound enough to live by.

> In the early years of the Second World War, when by coincidence he was a professor in Königsberg [Kant's home city], Konrad Lorenz used Darwin's idea about the formative role of the past to put the finishing touches to Kant. He argued that Kant's scepticism about what the world is really like was unjustified because the cognitive structure which enables us to know what the world is like had evolved through natural selection. The reason why our minds have this particular, and no other, cognitive structure ... must be that we have evolved and not flown in, so to speak, from outer space. Our cognitive structure has been selected by and, therefore, reflects or represents, the real world. (Munz, 1993).

So what is inborn in the individual has arisen by the accumulated actions of natural selection upon countless generations of ancestors. In technical bio-philosophical language, such inborn concepts (or, as we would now say on the basis of rather better genetic understanding, *propensities to form* concepts) are "... ontogenetically *a priori,* but phylogenetically *a posteriori*" [innate in the individual but experience-based in the species] (Lorenz 1977).Thus:

> The great resistance of the empiricists to innate knowledge is made irrelevant ... in the form of a more encompassing empiricism.
> (Campbell 1974 p.425).

Yet this positions us to think even more broadly, about the nature of knowledge itself. It is an outcome of the world's effects on us, both over evolutionary time via genetic selection and over the individual life-time by means of synaptic selection and subsequent learning:

> From the perspective of biology, knowledge ... is a form of self-reference.... The knower is part of the known, and has been shaped by what is known. The reflector reflects, more or less adequately, because it is itself part of what is being reflected. The biological perspective, therefore, provides an assurance that the reflector is adequate and also explains at the same time how it has been shaped by natural selection to be adequate"
> (Munz, 1993)

Again:

> ... the impressive order in nature is not, as has been claimed by idealistic philosophy, a product of our thinking and imagination ... , on the contrary, human thought itself is a product of the emerging order in nature.
> (Wuketits, 1984)

One more instance of EE's contribution to classic philosophical debates concerns the inveterate human disposition towards generalization and induction. Hume pointed out that induction can be justified only psychologically, not logically, but EE explains why we all have this psychological drive: living beings can only adapt to consistent features of the world, but it is imperative for their survival that they do so adapt. Conscious beings must therefore have the propensity to look for such consistencies.

In all the respects discussed in this section, EE – "Philosophical Darwinism" – has at last made sense of common sense!

Exploratory knowledge

If human understanding were limited to that which could be directly derived from innate concept-forming propensities, its rate of progress would be restricted to that at which DNA mutates – either the DNA which directly affects these propensities, or that which modifies rates of synapse-change. Individuals could learn from experience, but there is no evident mechanism by which their learning could be passed on faster than the mutation rate. (The assumption that it could be conveyed by inheritance would be an instance of Lamarck's mistake.) Yet actually we can not only learn, but pass on our learning, at rates many orders of magnitude faster than that of mutation!

Mainstream evolutionary epistemologists, such as Lorenz, Campbell and Vollmer, regarded each actual concept we form on the basis of our inherited propensities as an hypothesis about the world: "hypothetical realism" is the label given to the resultant epistemological stance. These concepts – these hypotheses – are, of course, challenged by our encounter with our surroundings and, where they prove unfruitful, either *we* shall be eliminated (if they have drastic short-term consequences) or we, as conscious beings, will eliminate *them* in favour of alternative hypotheses which we shall put in their turn to experiential test.

An even more eminent philosopher, Karl Popper, although only fully espousing Darwinism quite late in his intellectual life and never, I think, using the term "hypothetical realism", took what was effectively this idea forward to form the basis of his mature philosophy of sophisticated knowledge, and particularly of science. Although asserting the consistency of all knowledge-growth "from the amoeba to Einstein", and regularly referring to his viewpoint by the label "EE", Popper (e.g. 1972) had little concern with the biological history, or the mechanisms, of *natural* selection. His chief interest remained, as it had always been (Popper 1934/1959), the process of successive "conjectures and refutations" by which knowledge advances in mature human societies. This is an aspect of the contention, sustained throughout his writing, that in the matter of factual beliefs about the world the only certainty ever possible to us lies in falsification, not verification. Even in science, therefore, hypotheses can never be proved, but should instead be so stated that they make specific predictions vulnerable to challenge; their success, if these predictions are not refuted, is merely that they have for the time being evaded disproof. Of the other authors cited, only Campbell *passim* and Munz, in much but not all of his writing, have had in mind this conscious level of conjecture and potential refutation when they used the term "EE". From Darwin, through James and Lorenz to Vollmer (and many authors between) the earlier, biological processes, affecting the unconscious propensity to form basic concepts and the selection of the better ones by preferential survival, is what "EE" chiefly refers to. The respective usages are not unconnected, yet their emphases are very different. Each is a valid meaning of the term, but they should not be confused.

Nonetheless, despite their radically different speeds, the two modes of learning share the same logical structure. In neither is a model of the external world imposed on the living entity by that world. Unicellular and other simple organisms do not even have body systems which could learn this way, yet by natural selection they become increasingly adapted to the world over evolutionary time. Thus they may be regarded as having acquired information, and so "learned", about it. The process is not only entirely unconscious but entirely random. Yet the result is a form of learning. Consider a bacterial species in which a mutation occurs, affecting its resistance to an antibiotic. If the mutant is more susceptible, it will die particularly quickly; but if more resistant it will flourish, and gradually replace the earlier strain. Such development of antibiotic resistance is an instance of the "random mutation and selective retention" by which Donald Campbell characterizes all learning. Most of us would hesitate to regard such a mindless process as within the compass of epistemology, but the learning processes which *are* within that compass are considered to be contiguous with such primitive learning.

It may be tempting to suppose that animals with nervous systems, to which the concept of an epistemology begins to become applicable, can also acquire information receptively, by the action of the outside world upon the individual. However, such a concept tacitly assumes that the world consists of discrete bits of information, which could be picked up by passive response or painstaking observation. Yet, in Gerald Edelman's phrase, the world "is not labeled", and the radical evolutionary epistemologist contends that such passive information-acquisition never happens. Instead, the "labeling" is done by the organism, imposing its comprehension-patterns on the world which it experiences. Random "conjecture" and the only occasional avoidance of "refutation" is thus proposed in EE as the picture of all learning. Its rate is limited by that of mutation where there is no nervous system, and of synapse modification where there is one. In advanced organisms such as ourselves, where the effectiveness of a conjecture is consciously assessed, learning can be extremely rapid – though current neuroscience still assumes that synaptic modification is required. But at every level of sophistication the logical structure is considered to be that of active exploration, not the passive receipt of incident information.

Metaphysical theology

The crux of the arguments I have adduced to this point, for the reliability of the concepts which we human animals are predisposed to

form, is that we could only have survived if they are more or less right. Natural selection operates in terms of behaviour not thought processes, but the behaviours considered so far could only be advantageous if the concepts on which they are based are a good fit to the world. However, at the other extreme from concepts of space or animality are those adduced by metaphysical theologians – concepts, for instance, of:
- The supernatural
- Immaterial existences, "outside" space and time
- Creation from nothing
- Eschatological prediction.

Natural selection provides no test for the validity, or even the meaning, of concepts in these categories. I therefore suggest that their objective credibility score must be asymptotic to zero.

To move from those general headings to particular examples is to risk offending some readers, but the attempt to be clear requires that I follow the logic of the stance I am advocating. If the effect is to convince you that I am wrong, that will be fair enough! I cite therefore, as being so remote from the reality-checks of life on this earth that their very meaningfulness must be questioned, the concepts of:
- Paradise, hell and (if it is still to be considered!) limbo.
- Any substance or events "preceding" the Big Bang.

(I once heard an eminent philosophical theologian debate, in a lecture, why God "waited so long" within eternity – or, on the other hand, didn't "wait longer" – before creating the universe! St Augustine of Hippo, 15 centuries ago, adopted a much wiser stance.)
- A new Heaven and a new Earth ... and all high metaphysical constructions of the Easter event, such as that of Robert Russell (2006):

> If God chooses to act in a radically new way, as at Easter, then the 'laws' which reflect this will be radically new laws, not the ones we have now. [Easter is thus] the first instantiation of a new law of nature.

- Reification of "the soul", leading to what I have to regard as irrational stances on abortion and contraception, and a hideously unethical one on AIDS prevention
- The metaphysical positions underlying age-long conflicts about the Eucharist, priesthood and episcopacy, and salvation.

These examples are all from Christian thought – though the first two do not have to be. I refrain from commenting directly on other faiths, but am clear that some are just as guilty of inflated and dangerous metaphysics.

Considering my examples from the standpoint of the previous section and the two different emphases within EE, all these theological concepts are subject to intensive intellectual scrutiny in internal terms, assessing their coherence with other doctrines – which are often themselves similarly metaphysical. Nor would any doctrine take hold if it did not ring true to a significant group of believers. In this very weak sense they might be said to have been subject to an experiential, and hence perhaps Popperian test (c.f. Stanesby 1985). Clearly, however, they cannot be subject to experimental challenge: they make no predictions which can be refuted in a laboratory, or by observation within the world. Yet my criticism of them goes much deeper than that, to the fundamental nature of their purported truth-claims: they are not just saying untestable things about the world of time and space, *they are saying nothing at all about that world*. Yet it is to that world, and only that world, that are our minds adapted; only concepts with meaning in that world have come through the long, long selection process which justifies our credence. Thus metaphysical imaginings of the kind I have listed fall at the first hurdle – the hurdle brought to our attention by Darwin, James, Lorenz and Vollmer – well before the Popperian level of sophisticated testing can be brought to bear.

Of course, the *behaviours which metaphysical religious concepts induce* have been relentlessly tested for their contributions to survival, but my point is that *their truth-content* never has been and never can be. Now what is subject to natural selection is always behaviour, so it could properly be argued that there are no concepts at all whose truths have been tested directly. However, as the evolutionary perspective has made abundantly evident, concepts of the physical world lead to behaviours which will only be advantageous if the concepts are highly accurate – if the space-concepts of the ancestor we share with monkeys are not correct to within a very small percent it will miss the branch. From the standpoint of EE it is precisely for this reason that we can place very considerable trust in the validity of such concepts. In crucial contrast, the behaviours promoted by religions – social bonding, moral rules, mating practices, self-sacrifice for the good of the group – depend on the *strength* of the belief which promotes them, but *not on its truth*. Thus the survival of such beliefs provides no grounds for trusting their validity – for attributing to them even meaning, let alone truth.

Comparisons with the arts and mathematics

If this conclusion is felt too stark, can we reconsider it in light of other thought-categories which depart from empirical testability? What about the creative arts – painting, sculpture, music? Can they be tested? Well, the representational aspects of the visual arts can, but only very occasionally can a similar test be applied to music; and it is only in these representational aspects that these arts make claim to representing objective facts. The respects in which they are actually *art* are subjective, and directly or indirectly behavioural. There is no model here for claiming that what is essentially untestable may nonetheless embody objective, factual truth.

Mathematics might seem more promising (as briefly suggested by Mladen Turk 2003): wild flights of mathematical imagination may, decades or even centuries later, provide extraordinary insights into physical reality. Einstein's use of Riemannian geometry is a well-known example. Riemann's airy, non-Euclidean speculations were found, some 60 years afterwards, to embody important truths about the real world. Are theological speculations in any way analogous? The thought is tempting, but I fear it is not justified. The basics of mathematics are physical-world realities, the arithmetic and geometry of what we can see, touch and manipulate (Lakoff & Núñez 2000; see also Hitchcock, this volume). Even though the imaginative structures constructed upon them can be stupendous, the foundations are among the most directly survival-tested concepts considered at the start of this essay. ... Other-worldly concepts, by definition, have no equivalent foundations. Or, more accurately, they may well have such foundations in our minds, but an infinitely deep chasm is leapt in the claim that they may be extrapolated to apply outside this world. There is no equivalent chasm at the heart of mathematics.

Origins of religious concepts

We cannot know with certainty how any particular religious concept arose. There have been many proposals, and it seems likely that not all concepts arose the same way. One of the more widely-canvassed ideas (e.g. by Pascal Boyer 2001 and Justin Barrett 2004) concerns the concepts of supernatural agency. There is compelling evidence, from studies of children and primitive peoples, that human beings have a strong innate disposition to think in terms of living agents as the causes of unexpected events. It is argued that there is substantially greater adaptive advantage in being over-inclined to suspect such an agent than in being under-inclined, for living agents are more likely than non-living ones to pose unpredictable threats. When no such agent is found, after strenuous searching, the hypothesis of an agent which is living but invisible constitutes only a small extension from regular experience.

Another proposal, by Caspar Söling (2004) seems particularly applicable to the more metaphysical religious concepts. It is that they arise "when intuitive ontologies [concepts of being] conflict", each alone being unable to fit the whole experience: thus the observation of bodily death, in conflict with the sense of continuing self-hood, gives rise to the idea of "soul".

A third recent suggestion, by David Sloan Wilson (2002) is of a different kind. Wilson holds to the concept of "group selection" as a force in evolution, and contends that the mutual support of members of a religious group provides them with competitive advantage. With or without the theory of group selection, the general idea that the behavioural consequences of religious beliefs must have been, on balance, beneficial has been around a long time and is widely accepted – societal bonding and moral codes being both reinforced by ritual as well as belief. The metaphysical ideas to which the group gives credence are of negligible account in themselves; it is the members' mutually supportive behaviour which matters. Peter Munz, who had argued fairly similarly to Wilson somewhat earlier (1985, 1993), even proposed that bonding was stronger when the knowledge-claims on which it was based were false:

> True knowledge makes a poor bonding principle, because sooner or later everyone will embrace it. ... But while there are few true theories, there is an infinite number of false theories. Each particular false theory divides the people who embrace it from all those who ... embrace a different kind of falsehood. Hence there has always been a presumption in favour of the survival of false knowledge. (1993 p.3)

These samples of current thinking about the driving forces behind religious ideas scarcely inspire one with confidence about those ideas being right. Nevertheless, Söling (2004) and van Huyssteen (2006) both claim that (to quote Söling's words) "the way religious concepts arise says nothing about their validity". This is certainly true of the chance historical paths by which concepts come to be formulated – which individual human being, in which circumstance, first propounded the concept being considered. And it is probably acceptable in relation to the Boyer/Barrett and the Wilson proposals, as well as to Söling's own. But Munz's argument, in the quotation above, actually goes further and suggests that the more strongly held the belief, the more probable it is to be untrue!

The challenge of EE

I finally return to the question with which I began, and ask whether van Huyssteen, despite his courage in recognizing that EE presents a challenge to theology, has fully grasped the severity of that challenge.

We have seen repeatedly that, if there has been and can be no selective check on the purported truth-content of a concept there is no sound basis for accepting it as true.

Perhaps you feel that EE, as thus recounted, is nothing new – just naturalism writ large? I do not object to that view, but would contend that the persuasive nature of the Darwinian approach makes its naturalistic challenge particularly hard to avoid.

From this challenge I conclude that the metaphysical concepts of theology must be one of the following:
1) Special divine injections (an ancient view reinvoked, with specific reference to EE, by Tomáš Hančil & Margarete Ziemer, 2000)
2) Indirect products of a top-down influence from God on evolution, of the sort proposed by the late Arthur Peacocke (1979, 1993)
3) Metaphorical clutchings after an impossible aspiration toward objective concepts
4) Meaningless.

Both (1) and (2) are logically possible, but they must be accepted as acts of faith. There can be no evidence that they are the case, because they make no falsifiable predictions. For most people of scientific temper they breach William of Occam's golden rule, paraphrasable as:

> Don't adduce mechanisms you can manage without.

In these terms Donald Campbell (1974) proclaims the *credo* of EE:

> At no stage has there been any transfusion of knowledge from the outside, nor of mechanisms of knowing, nor of fundamental certainties.

Of our listed possibilities (1), in particular, smacks of wishful thinking. Tomas Hančil & Margarete Ziemer, after an admirable four-page exposition of EE, "widen the presupposition for their investigation by allowing for divine influence" (2000 p15). Van Huyssteen himself (2006 p.105) considers that this approach, while claiming to deploy EE, takes:

> one step too far by ... invoking ... God in the midst of an epistemological argument.

Any fundamentally scientific thinker would agree: This is the introduction of a *"Deus ex machina"* – it is not part of science, and not compatible with it.

However, van Huyssteen offers no substantive alternative – not even (2) which, though not itself part of science either, *is* science-compatible. Instead van Huyssteen stresses:

> the radical embodiment of our human condition" making us "incapable of envisaging entities that cannot be construed from what we know of the material world (283)

Right! And he quotes Ian Tattersall (1998):

> We might do well to look on the inadequacy of our concepts of God as the truest mirror of those limitations that define our condition.
> (Cited by van Huyssteen, 2006 p 283)

Van Huyssteen is almost there! But he doesn't quite grasp the nettle. Instead he asserts:

> ... the theory of evolution by natural selection cannot offer an adequate explanation for beliefs that far transcend their biological origins (89)

Wrong! In the previous section we saw that it can and does! More sympathetically, in the same spirit as Tattersall:

> We continue to imagine God in our own image simply because ... we are unable to do otherwise. Importantly, however, this does not imply the illusory character or non-existence of God, but might actually reveal the only intellectually satisfying way to talk about God ... (283)

The first sentence has obvious truth, though some of the characteristics we attribute to God are rather remote from our own image: few of us, for instance, are three persons, and if we are it is considered a consequence of psychopathology not divinity. The second sentence, however, is the nub. While its second clause may well be true, only faith can uphold the first one.

****

We have seen that the insights of EE confront us with the absolute & total embodiment of mind. Even St. Paul acknowledged that human beings are "of the earth, earthy" (1Cor. 15: 47) and cannot escape those shackles. The implication is that, to avoid conclusion (4) of the previous page, the biologically-founded thinker must accept (3); the propositions of metaphysical theology have psychological value but no objective validity.

Many, from Feuerbach on, have decided on such grounds that the idea of God itself, and the associated dispositions to reverence, awe and worship, must be eschewed. In the current decade Boyer (2001), Richard Dawkins (2006) and Sam Harris (2007) have, temperately or intemperately, asserted such a view. I do not. For me, such extreme negations are both logically and emotionally unacceptable. The Darwinian approach does nothing to contradict the concept of a first cause and indeed, as Aubrey Moore (1893) saw clearly over a century ago, it invites that of a continual, immanently sustaining one. And if we sought to reject every emotion, every apparently-spiritual perception, because we can recognize its evolutionary basis, we would allow no space for duty, sacrifice or heroism, and regard romantic love with derision. In any case, all these, too, are sometimes the subjects of attempted rationalizations in terms of a world beyond experience; but when the rationalizations prove unsatisfactory, we do not ban the emotions from our lives. My argument remains an Occamite one – that we should accept the a-rational foundations of all the deep dispositions which are of value to us, but not build upon them intellectual edifices which these foundations cannot support.

On this basis I conclude that metaphysically inclined theologians must:
- make no claim to realism
- and attempt no positive statements about non-physical worlds or states of being .....
- or about the character of the Divine.

Instead, like the elderly Thomas Aquinas, they should:
- confront the *mysterium tremendum* with humility and wordless awe,
- follow the *Via Negativa*,
- immerse themselves in the *Cloud of the Unknowing*,
- and continually heed "the silence underlying sound".

If they really believe that deity is ineffable, should this present a difficulty?

The Darwinian approach to theology thus leads me to urge a strenuous opposition to dogmatism of any form – fundamentalist or hierarchical, Christian, Judaic, Islamic or Hindu. Though my route to that stance may have been unorthodox, in the current world should not any approach which leads to this conclusion be welcomed and encouraged?

References

Barlow, H.B. & Mollon, J.D. 1982. *The Senses*, Cambridge: University Press.
Barrett, J.L. 2004. *Why would anyone believe in God?* Walnut Creek, CA: Altamira.
Boyer, P., *Religion Explained*, New York: Basic Books, 2001
Brindley, G.S. 1960. *Physiology of the Retina and Visual Pathway*, London: Edward Arnold.
Campbell, D.T. 1974. Evolutionary Epistemology, in P.A. Schilpp (ed) *The Philosophy of Karl Popper*, La Salle, IL: Open Court.
Changeux, J P. 1986. *Neuronal Man* (trans), Oxford: University Press.
Darwin, C. 1871. *The Descent of Man,* London: John Murray.
Dawkins, R. 2006. *The God delusion,* London: Bantam.
Edelman, G.M. 1989. *Neural Darwinism.* Oxford: University Press.
Farley, E. 1990. *Good and Evil: Interpreting a Human Condition*, Minneapolis, MN: Fortress.
Gruber, H.E. 1974. *Darwin on Man,* New York: Dutton.
Hančil, T. & Ziemer, M. 2000. Evolutionary Epistemology as a New Challenge in the Dialogue between Theology and Science, *CTNS Bulletin*, 20/3.
Harris, S. 2007. *Letter to a Christian Nation,* London: Bantam.
Haught, J.S. 2000. *God after Darwin,* Boulder, CO: Westview.
Hecht, S., Shlaer, S. & Pirenne, M. 1942. Energy, quanta and vision, *J. Gen. Physiol.*, 25, 819.
Hefner, P. 1993. *The Human Factor,* Minneapolis, MN: Fortress.
James, W. 1890. *Principles of Psychology*, New York: Henry Holt.
Kaufman, G.D. 2004. *In the Beginning ... Creativity,* Minneapolis, MN: Fortress.
Lakoff, G. & Núñez, R.E. 2000. *Where Mathematics Comes From: How the Embodied Mind Brings Mathematics into Being.* New York: Basic Books.
Lorenz, K. 1941. Kants Lehre vom Apriorischen im Lichte gegenwärtiger Biologie, *Blätter für Deutsche Philosophie,* 15, 94.
—. 1973. *Die Rückseite des Spiegels,* Berlin: Piper. [Translated as *Behind the Mirror*, New York: Harcourt Brace, 1977]
Moore, A. 1891. The Christian Doctrine of God, in Gore, C. (ed) *Lux Mundi,* London: John Murray.
Munz, P. 1985. *Our Knowledge of the Growth of Knowledge: Popper or Wittgenstein?* London: Routledge & Kegan Paul.
—. 1993. *Philosophical Darwinism,* London: Routledge.

Peacocke, A. 1979. *Creation and the World of Science,* Oxford: University Press [pbk reprint 2004]
—. 1993. *Theology for a Scientific Age,* London: SCM.
Plotkin, H. 1995. *Darwin Machines and the Nature of Knowledge,* London: Penguin. [Originally published Cambridge, MA: Harvard, 1993.]
Popper, K.R. 1959. *The logic of scientific discovery,* London: Hutchinson. [Originally published as *Logik der Forschung.* Vienna: Springer, 1934.]
—. 1972. *Objective Knowledge: An Evolutionary Approach.* Oxford: Clarendon
Russell, R.J. 2006. *Cosmology, Evolution, and Resurrection Hope.* Kitchener, Ont: Pandora.
Simpson, G.G. 1963. Biology and the Nature of Science, *Science* 139, 81-88.
Simmel, G. 1895. Über eine Bexiehung der Selectionslehre zur Erkenntnistheorie, *Archiv für systematische Philosphie* 1.1.
Söling, C. 2004. The Instinct for God, *ESSSAT News* 14.2, 11-12.
Spencer, H. 1897. *Principles of Psychology,* New York: Appleton, [1st ed. 1855]
Stanesby, D. 1985. *Science, Reason and Religion,* London: Croom Helm.
Tattersall, I. 1998. *Becoming Human: Evolution and Human Uniqueness,* New York: Harcourt Brace.
Turk, M. 2003. On Pattern Recognition ..., in W.B. Drees (ed) *Is Nature ever Evil?* London: Routledge.
Van Huyssteen, W. 1999. *Duet or Duel?.* London: SCM
—. 2006. *Alone in the World,* Grand Rapids, MI: Eerdmans.
Vollmer, G. 1984. *Mesocosm & Objective Knowledge,* in Wuketits (1984).
Von Uexküll, J. 1928. *Theoretische Biologie,* Berlin: Springer.
Wilson, D.S. 2002 *Darwin's Cathedral.* Chicago: University Press.
Wuketits, F.M. (ed) 1984 *Concepts & Approaches in Evolutionary Epistemology.* Dordrecht: Reidel.

Chapter Seven

Reply to Professor Spurway

Derek Stanesby

Revd Dr Derek Stanesby took a first degree in philosophy at the University of Leeds in 1956, and a PhD from Manchester in 1985. He trained for the Church of England ministry at the College of the Resurrection, Mirfield, from which he was ordained in 1959. He was a parish priest for 27 years, then Canon of St George's Chapel, Windsor for 12, before retiring in 1997. While at Windsor he ran a series of highly-regarded Consultations on Science & Religion, as well as chairing the Science & Religion Forum. In 1985 he published "Science, Reason and Religion", an exposition of the value of Karl Popper's thought for the science/religion debate.

Derek Stanesby is also deeply versed in the writings of Peter Munz. His comments here demonstrate readings of both Popper and Munz almost diametrically opposite to those of the previous paper.

Spurway's approach

I find it difficult to unravel all the crossed threads of Professor Spurway's argument but will attempt to pick out his line of thought in order to offer an appraisal. He suggests that "Darwinism" is a good starting point for epistemology – the theory of knowledge. "Darwinism" in this context is expressed in "the discipline of Evolutionary Epistemology (EE)". He continues:

> The basic contention of EE is that not only our bodies but our brains, and not only our physical brains but the kind of concept they can form, are all entirely products of natural selection.

EE trades particularly on Lorenz's observation that our sense organs, having been selected, must be adapted and therefore trustworthy sources of knowledge. The mind is not a *tabula rasa* as envisaged by Locke, but it has evolved as a result of natural selection to pick up correct information about our immediate world without which we would not have survived. Spurway extends this argument to social judgments and dispositions such as the ability to learn a language. "Not only our perceptions but also our *con*ceptions must be pretty accurate" for our survival. He quotes Vollmer in support, "human cognition arose according to natural law" and concludes that our knowledge "is an outcome of the world's effects on us." He comes close to justifying induction because of our common-sense propensity to look for regularities in the world.

Spurway next moves from the inherited, adapted products of natural selection to Popper's consciously articulated process of conjecture and refutation. Thus, he maintains, we have two modes of learning: the gradual adaptation of our physical organs (including our brains) through the inordinately slow process of evolutionary natural selection, and the much faster route of consciously articulated Popperian conjecture and refutation.

Spurway finally claims that in contrast to the reliability of concepts, which through natural selection are "a good fit to the world", there is a certain set of concepts adduced by what he calls "metaphysical theologians" which have no objective credibility because:

> ... natural selection provides no test for the validity or even the meaning, of concepts in these categories.

Further:

> they cannot be subjected to experimental challenge: they make no predictions which can be refuted in a laboratory, or by observation within the world ... they are not just saying untestable things about the world of time and space, *they are saying nothing at all about that world.* Yet it is to that world, and only that world, that our minds are adapted; only concepts with meaning in that world have come through the long, long selection process which justifies our credence.

Quite how such religious ideas arose is unclear, according to Spurway, but given his opening assumption about concept formation they are part and parcel of all concepts which are products of natural selection. He does not dispute van Huyssteen's contention that EE explains "how God-concepts arose", that religion is natural to human kind, but he claims that they cannot be trusted, they have no truth value, they are literally non-sense and therefore nonsense. They have psychological value (not least in the binding and bonding of communities) but no objective validity. His advice to "metaphysical theologians" is to be silent, say nothing and immerse themselves in the *Cloud of Unknowing.*

Further refinement of Spurway's argument gives us three main propositions.

1. All our concepts are entirely the products of natural selection. "Evolutionary Epistemology claims that ... human cognition arose according to natural laws" (Vollmer). All this has taken place over aeons of evolutionary time and proceeds at the rate of gene mutation. The introduction of the term "hypothetical realism" to label the process of "random mutation and selective retention" (Campbell) points to the fact that it is the noumenal, or real world to which our senses become attuned. It is an unconscious and entirely random process but it implies that the world instructs us and gives us knowledge.

2. The Popperian process of consciously articulated human conjecture and refutation follows the same logic as above but it advances knowledge far more quickly.

3. Concepts which relate directly to the world are reliable because they have passed through the sieve of natural selection but metaphysical religious concepts must be rejected because they do not relate to the world and have not been naturally selected.

There is much confusion here. No link between propositions 1 and 2 is provided and they appear to be unrelated. This is because the term "evolutionary epistemology" is used in two distinct senses: one as a paradigm to refer to Popper's theory of knowledge growth and the other as a biological process. Proposition 3 makes no sense because "metaphysical religious concepts", like all concepts, are purported to be the products of natural selection yet "natural selection provides no test for the validity, or even the meaning, of concepts in these categories". Confusion arises here because "validity" applies to linguistically formulated hypotheses and not to the products of natural selection which are non verbal. This curious conflation of ideas results from the confusion of natural selection with the development of culture. It is often said that there are three phases of evolution, physical, biological and cultural. This is misleading because "culture" has not evolved according to natural laws, it develops within communities and societies and becomes increasingly dependent on language. Animal learning through natural selection and biological adaptation should not be confused with linguistically formulated human learning or the growth of knowledge. The former is an unconscious process involving genetic inheritance, the latter is a conscious process involving education. Knowledge, traditions, beliefs are not inherited but passed on from generation to generation. Hence the enormous significance of educational programmes in advanced societies. Religion has not evolved in a biological sense, it may be widespread but it is not "natural".

Religion has developed within cultures: its development has nothing to do with Darwinism. The appeal of the contrary view, that belief in God is a matter of nature rather than nurture, that "belief in God is natural", as one Oxford theologian has put it, is evidenced by the allocation of £1.9 million to research this issue in the University of Oxford. (*Times,* 19 February 2008)[1]. The allocation of such vast resources to such an enterprise is doubtless motivated by an attempt to establish religious belief on firmer "scientific" foundations, in the face of increasing hostility towards religious belief by the likes of Richard Dawkins, but no amount of research can establish religion as a product of evolution by natural selection, as will become apparent. Much the same is true of the wide-ranging conjectures of natural science which is also a cultural phenomenon.

Popper's refutation of positivism

The root of Spurway's confusion derives from his adoption of the labels "Darwinism" and "Evolutionary Epistemology". The term "evolutionary epistemology" was first coined by D.T. Campbell and refers, particularly in his famous contribution in the Schilpp volume *The Philosophy of Karl Popper* (Campbell 1974), to Popper's epistemology. Popper refers to an earlier version of this in his 1972 book *Objective knowledge* in which he emphasises that the *logic* of Darwinian natural selection is identical to the *logic* of the growth of knowledge – from the amoeba to Einstein. "The method of trial and the elimination of errors belongs to the *logic of the situation".* Popper developed his epoch-making alternative to conventional positivistic epistemology, in his *Logik der Forschung* (1934) and subsequent writings without reference to or reliance on Darwin. Popper's Darwinism has nothing to do with the alleged Darwinism of biological evolutionary epistemology or evolutionary psychology; in fact they are diametrically opposed to each other. Spurway unfortunately confuses and conflates these two irreconcilable paradigms. The suggestion that Popper "only fully espousing Darwinism quite late in his intellectual life" took the idea of evolutionary epistemology "forward to form his mature philosophy of sophisticated knowledge, and particularly of science" simply will not pass scrutiny. Popper's epistemology is in no way foundational and does not depend on a natural selection theory of cognition as a starting point. He develops Kant's *a*

---

[1] The directors of this project are Prof Roger Trigg (Chapter Eight herein) and Dr Justin Barrett, whose work is cited in Chapters Two and Six. Ed.

priori in terms of innate ideas without reference to Lorenz. His whole epistemology is ruthlessly critical of induction which he dismisses as part of positivist epistemology. He never referred to falsification in terms of certainty – he made a clear distinction between logical and methodological falsification. Most importantly, Popper's epistemology is entirely in the realm of linguistically formulated theories which can be criticised, provisionally accepted or rejected and it is only by analogy that it can be likened to Darwinian natural selection.

Spurway correctly identifies the idea of biological evolutionary epistemology with Lorenz's observation regarding the selection and adaptation of our sense organs as trustworthy sources of knowledge. As Popper (1972) comments:

> If the process of adjustment has gone on long enough then the speed, finesse and complexity of the adjustment may strike us as miraculous.

– a point made by Spurway with regard to vision and hearing for example. However, Lorenz was concerned initially with animal behaviour and his observation in this regard is certainly valid for the kind of information with which paramecia and mallard ducklings come into the world and which allows the monkey to grasp the branch. It is this "primitive" or instinctual brain which *homo sapiens* has inherited and which equips us with the *a priori* Kantian categories of space, time and causality. In this sense our primitive cognitive structures are pretty accurate at representing our immediate world to us. However, a rough guide to the human brain describes a far more complex cognitive structure with the integration of the cerebral cortex with the primitive mammalian brain. Neuroscience has shown that the human brain is not a single organ capable, with a few exceptions, of responding unequivocally to stimuli. It registers colour, position, size, time, location, shape, language and many other inputs in separate parts of the brain and thus creates a binding problem which has to be resolved before a single representation can emerge. Human cognition is thus of a completely different order from instinctive animal adaptation to the local environment. Lorenz's insight came to a halt with the emergence of this large and complex human brain. What Popper demonstrated was that just as biological organisms can be considered as proposals made to the environment and then exposed to natural selection, so linguistically formulated theories are exposed to falsification by empirical testing, or more generally by criticism. As Peter Munz (1993) has put it, "organisms are embodied theories and theories are disembodied organisms". Popper's epoch-making alternative to conventional positivism consisted in the fact that he saw a complete continuity, in terms of problem solving, from the amoeba to Einstein. If the amoeba fails to solve its problem it is eliminated whereas with Einstein, his theories die in his stead. Campbell expressed this in terms of chance mutation and selective retention both in biological evolution and in the Popperian paradigm of the evolution or growth of knowledge. There is thus a world of difference between the Popperian paradigm and evolutionary epistemology as developed by Vollmer and others and taken on board by Spurway, which holds that human knowledge growth is of the same kind as animal learning, that both are "an outcome of the world's effects on us" and therefore determined by our environment. The lure of positivism is evident here because of the desire for a trustworthy starting point or reliable foundation for our knowledge claims. Old sense experience has been replaced by new natural selection; there is nothing in the mind that has not been naturally selected to be there. It is to such naïve positivism that Popper was vehemently opposed.

Culture and language

The question arises, how was the move made from natural selection in biology to cultural development and the growth of knowledge in science and the excursions of the human mind in literature, poetry and religion? The brief answer is through the evolution and development of language. The evolution of language is tied up with the evolution of consciousness. "Consciousness" is a portmanteau word covering a wide range of conditions from primitive stimulus and response to human articulate self-conscious awareness. There is an evolutionary scale here and clearly a world of difference between reacting to the dentist's drill and reflecting on the alpha and omega of existence. Just as there are different levels of consciousness so there are different levels and functions of language; the two are interdependent. Primitive levels of consciousness are evident in animals as are primitive functions of language, in terms of its symptomatic and signalling functions. The development of the descriptive and argumentative functions of language marks the emergence of man from the animal kingdom (see Stanesby 1985 pp.59 ff). Munz develops these ideas by suggesting that the modification of animal language into the speculative or creative language that is characteristically human language is a move from what he calls 2-dimensional language to 3-dimensional language and the development of a general purpose mind.[2] It is a move from talk about matters of fact to talk expressing hope, fear, possibility, past, future, heaven, hell, God and the devil and more. It allows excursions

---

[2] See also Deane-Drummond, Chapter Two of this volume. Ed.

of thought from the Bible to Homer to Shakespeare and Milton to the modern novel; from Galileo to Newton to Einstein; from the atoms of Democritus to quantum theory to string theory; from one world to many worlds and so on. This is a world away from Spurway's contention that the kinds of concepts we can form are "entirely the products of natural selection".

Munz makes an astute and crucial observation by identifying the gap between neuronal activity in the brain and the verbal labelling of what Damasio (1994, 1999) calls "somatic markers". Somatic markers, or to use an older term, "qualia", are the inchoate feels and feelings experienced by an organism with a degree of consciousness, which become more intense with a more developed self conscious awareness. These inchoate experiences do not come with verbal labels attached to them. The neurons are silent; they do not have words glued to them. In other words there cannot be a direct causal path from the physics and chemistry of neuronal and synaptic activity in the brain to the subjectively experienced and verbalised states of mind which constitute our consciousness. Munz proposes, therefore, that we understand consciousness as a composite phenomenon consisting of three layers – neurons, somatic markers and explicit states of mind. The connection between the first two layers is causal and can be described in terms of physics and chemistry, and the relation between the second and third layers is *interpretive* and therefore hypothetical and uncertain and is located in cultural development. He argues, persuasively I think, that this applies to all psychological statements about pain or pleasure or colour or whatever. Language is generated in separate parts of the brain (the Broca and Wernicke areas) from regions associated with our corporeal sensations. There is a subtle interplay between language use and the development of self-conscious awareness. We form hypotheses and make guesses in attempting to talk about our inner lives and in so doing we become more self aware. Although we have evolved an innate disposition regarding deep language structure (Chomsky) our actual language use is developed in communities or societies and is essential to cultural development. We must distinguish here between the evolution of our speech apparatus, in terms of the larynx and voice production, and the neuronal activity that generates it, and the actual *use* of language. Quite how words are found to interpret somatic markers, moods and feelings remains a mystery and possibly one that will not yield an answer, but we can get a good idea of how people have dealt with the problem by examining the development of language use in literature. Having invoked Popper in terms of problem solving and the growth of knowledge by trial and error, we must invoke Wittgenstein in terms of what he calls *language games* or *forms of life*. Wittgenstein, who in the *Tractatus* (1922) identified linguistic labels with simple facts about the world, realised his mistake and in examining the use of language realised that there is no one-to-one relationship between language and the world. Subsequently, in the *Philosophical Investigations* (1953) he identified *meaning* with *use* thus taking the sting out of the logical positivists' accusations concerning *meaninglessness*. Language is manufactured and developed in a community and follows agreed conventions and rules. In our language games we experiment with language to describe feelings, moods and ideas. Hence the possibility of literature, poetry, science and religion. Thus the attempt to lace the cultural development of language into the causal network of evolutionary epistemology or evolutionary psychology is doomed to failure. Despite the complexity of the issues raised here, with regard to the integrated development of the brain, consciousness and language, the distinction between biological evolution and cultural development is essential to our understanding of the human mind and its extraordinary capacities.

This is the briefest summary of a fascinating thesis developed by Munz in his *Critique of Impure Reason* (1999) and in his last book *Beyond Wittgenstein's Poker* (2004) in which he is equally devastating in his criticism of evolutionary psychology. The popularity of this extraordinary thesis, like that of evolutionary epistemology, can be put down to the misleading impression that it is evolutionary and Darwinian. It is neither. It provides a flawed and inadequate account of how the mind works by postulating the evolution of domain specific modules in the brain which mirror specific features of the world, implying that we learn by instruction (positivism) rather than by selection (Darwinian). Neither evolutionary epistemology nor evolutionary psychology allow for the development of an all-purpose mind which is capable of going *beyond* the information given. Stephen Mithen (Chapter One of this book) is aware of this problem and tries to get round it by introducing yet another notion, *cognitive fluidity*. But that is unsatisfactory too, because this purported fluidity can only integrate the given, biologically determined, domains of the brain and likewise cannot produce a mind that is capable of imaginative conjectures which take us beyond the fixtures of natural selection. Munz's argument against both evolutionary epistemology and evolutionary psychology is that they represent a return to the old positivism according to which the world instructs us. This leaves no room for imaginative flights of fancy which are essential to the growth of knowledge. Three dimensional human language, on the contrary, allows us to go beyond the information given by forming imaginative hypotheses, guesses and conjectures. In biological natural selection, mistakes are made and eliminated by the environment, whereas with human imaginative conjectures mistakes are made and eliminated by criticism. The criticism in natural science takes the form of empirical testing or falsification and in other areas in terms of the problem situation under discussion (see Stanesby 1985 pp.103ff). The move from the positivist identification of rationality with justification, proof and certainty to criticism is at the heart of Popper's epistemology. The desire to establish knowledge on firm foundations is at the root of all positivist epistemology, but it is doomed to failure. All our knowledge is provisional. Browning's famous line, "Ah, but a man's reach must exceed his grasp, Or what's a heaven for?" is not only a battle cry in support of human aspiration and endeavour but also a recognition that we can take risks with ideas that enable us to reach beyond our grasp – that is, beyond the information given.

The dismissal of theology

Spurway invites us to consider theology in the light of evolutionary epistemology, but having critically disposed of this thesis as an explanation of the genesis and growth of human knowledge it is appropriate to consider his underlying contention which appears to be that religion is meaningless nonsense because it is not science:

> ... theological concepts ...cannot be subject to experimental challenge: they make no predictions which can be refuted in a laboratory, or by observation within the world.

His double whammy "metaphysical religious concepts" implies that his fundamental adversary is "metaphysics". This was the main contention of the totally discredited logical positivists of the early 20th century. In claiming that "only concepts with meaning in the world have come through the long, long selection process which justifies our credence", Spurway falls into the trap that ensnares all positivists – the resort to metaphysical assertions in order to eliminate metaphysics. Shades here of the metaphysical verifiability criterion of meaning, the flagship of positivism, which foundered on its own self proclaimed principle. It would be tedious to go through the arguments that eventually led to the demise of positivism, but one of the fundamental reasons for rejecting positivist epistemology is that its rejection of any form of metaphysics necessarily results in the rejection of natural science. In throwing out the bathwater of metaphysics, Spurway unwittingly throws out the baby of natural science as well (see Suppe 1977; Stanesby 1985). Spurway's list of no go areas lumps together the dubious notions of paradise, hell and limbo with any attempt to investigate "events preceding the big bang". Cosmologists, including the Astronomer Royal and President of the Royal Society, who venture into this forbidden land, beware! (See, for instance, Rees 2001). Presumably the academically respectable area of the philosophy of religion, explored by both theists and atheists, is also forbidden territory because Spurway's earlier strictures regarding "metaphysical theology" imply that we are not even able to formulate meaningful questions concerning the existence of God. It is one thing "to urge a strenuous opposition to dogmatism of any form – fundamentalist or hierarchical" but entirely other to demand the elimination of all metaphysical conjectures and attempts to understand the world in theological terms. The rejection of anything beyond science (meta physics) places us in the unimaginable bleak world of logical positivism and scientism.

My critical appraisal of Neil Spurway's paper is an attempt to deal with his arguments and not his intentions. I agree with him that a good bit of nonsense, in the general meaning of that term, is found in much theological speculation and that in many respects the less said by religious claimants the better. It is a truism to add that much nonsense is also found in other areas of human discourse, not least in philosophy. The laudable aim of the positivists from Vienna early last century was to purge the human mind of much metaphysical verbiage and provide a secure foundation on which truth could flourish. Their great mission failed, not least because they did not understand the nature of the human mind. We still do not understand it, but the revived interest brought about by neuroscience, with its insights regarding the structure and functioning of the human brain, along with critical philosophies of mind, have sharpened our awareness of the problems involved. The root of my objections to Spurway's argument concerns what Munz refers to as the widespread but "slapdash assumption", that human language is directly related to neuronal events, that "human neuronal systems have semantic properties" (Munz 1993). Linguistically formulated concepts are not sifted out at the level of natural selection. A great divide exists between the process of Darwinian natural selection, which fixes or adapts the organism to its environment, and the linguistically endowed human mind which takes us out and away from the world in which it is embodied. It is in this world of speculative thought that science and religion, among other things, flourish, and in which knowledge grows along with superstition and wild ideas, to which our sharpest critical faculties need to be applied. It is to this world of linguistically formulated theories and ideas that the Popperian epistemology of conjecture and refutation applies.

A final word from Popper (1982) offers hope for those of us who are prepared to take risks with our ideas and not close down the aspirations of the marvellous minds with which we humans have been endowed:

> The proper aspiration of a metaphysician, I am inclined to say, is to gather all the true aspects of the world (and not merely its scientific aspects) into a unifying picture which may enlighten him and others, and which one day may become part of a more comprehensive picture, a better picture, a truer picture.

References

Campbell, D.T. 1974. Evolutionary Epistemology, in P.A. Schilpp (ed) *The Philosophy of Karl Popper*, La Salle, IL: Open Court.

Damasio, A.R. 1996 [1994]. *Descartes' Error; Emotion, Reason and the Human Brain.* London: Papermac.

—. 2000 [1999]. *The Feeling of What Happens: Body, Emotion and the Making of Consciousness.* London: Vintage.

Munz, P. 1993. *Philosophical Darwinism: On the Origin of Knowledge by Means of Natural Selection.* London: Routledge.

—. 1999. *Critique of Impure Reason: An Essay on Neurons, Somatic Markers and Consciousness.* London: Praeger.

—. 2004. *Beyond Wittgenstein's Poker: New Light on Popper and Wittgenstein.* Aldershot: Ashgate.

Popper, K.R. 1959 [1934]. *The logic of scientific discovery*, London: Hutchinson.

—. 1972. *Objective Knowledge: An Evolutionary Approach.* Oxford: Clarendon

—. 1982. *Quantum Theory and the Schism in Physics.* London: Hutchinson.

Rees, M. 2001. *Our Cosmic Habitat.* Princeton: University Press.

Stanesby, D. 1985. *Science, Reason and Religion.* London: Croom Helm

Suppe, F. 1977. *The Structure of Scientific Theories.* Chicago: University of Illinois Press.

Wittgenstein, L. 1922. *Tractatus Logico-Philosophicus* (trans. C.K. Ogden). London: Routledge & Kegan Paul.

—. 1953. *Philosophical Investigations* (trans. G.E.M. Anscombe). Oxford: Blackwell.

CHAPTER EIGHT

ARE WE GHOSTS OR MACHINES?

ROGER TRIGG

Roger Trigg studied classics and philosophy at Oxford, then took a DPhil on the concept of pain, supervised by Professor Gilbert Ryle. He culminated his academic career as Professor of Philosophy in the University of Warwick, retiring to Emeritus status in 2007. He is now Senior Research Fellow at the Ian Ramsey Centre, Oxford, co-directing a major project on "Empirical Expansion in Cognitive Science of Religion and Theology", funded by the Templeton Foundation. Previously he has been President, at different times, of the British Society for the Philosophy of Religion, the Mind Association and (currently) the European Society for Philosophy of Religion. His many books include "Reality at Risk: a Defence of Realism in Philosophy and the Sciences" (1980, 1989), "Ideas of Human Nature" (1988, 1999) and "Religion in Public Life" (2007).

In this extended version of his conference paper, Prof Trigg upholds the traditional "dualist" view of mind or spirit as a separate entity, over against the material body, and argues that without this, the equally traditional theistic concept of God as a being separate from the physical world becomes, in its turn, hard to maintain.

Evolution and the Mind

The apparent success of neo-Darwinism in providing explanations for biological evolution has been coupled with recent progress in understanding the composition of the human genome. The result has been ever greater emphasis on our embodiment. We are seen as animals, complex certainly, but nevertheless not very different in kind from other species that have evolved over the millennia. Yet this goes against our self-understanding as humans, and the personal understanding of each of us of the importance of our own consciousness. We all feel that, to some extent, our future lies in our own hands. We feel free to reason, and to act accordingly. We feel that we continue as the same person through tremendous changes to our bodies. A person of ninety could remember experiences she herself had at three, and feel the same person – indeed may still be affected by such experiences. In other words, we are more than our bodies, and even different from them. This becomes a particularly strong intuition, when someone feels constrained, and even imprisoned, by physical disability. The issue becomes even more acute when we face death, and wonder if the end of the body means the end of me or of you. These common thoughts form the starting point for philosophers' concern about the connection with the body of whatever makes me the person I am. To put it another way, does personal identity demand physical continuity? Is the self necessarily embodied? Indeed, is it wholly constituted by the body (or the brain)?

One response to these issues is that of the dualist, who considers that there are two components to a human being, the body and something else. What the second item may be, continuing perhaps even beyond death, is called by various names. To some it is merely the mind, and, indeed, the nature of consciousness is often thought to be an intractable problem. Some philosophers consider that it will forever remain mysterious. Others talk of the "spirit" or "soul" and often imply that it is a detachable entity, entering us at some point before or at birth, and departing at death. To some, its presence marks the difference between a human and an animal.

It is hardly surprising that a tough-minded scientist may be impatient with such talk of immaterial entities outside the scope of science. There is the perennial issue of how an immaterial entity can interact with the physical one. How does, for instance, a mind control the body? Scientists can look at neural activity, but they cannot observe the interaction of that with something which by definition remains unobservable to the scientist. It is easy to conclude that such metaphysical entities do not exist, and are imaginary. Empirical science, it may be said, has to rule them out, because by definition they cannot be observed or measured. Yet, a rejoinder may be that such science is based on human experience. It is the individual experience of each one of us that the content of our consciousness can and does influence our body. How it is mediated by the brain may be impossible to determine, but that does not mean that only the brain is involved. None of us naturally thinks that we are only our brains. The most reductionist of philosophers who say that the mind is merely the brain, still have to admit there is what they call derisively "folk psychology". We always talk of our thoughts and our feelings, our emotions and desires. We do not for a moment ordinarily consider that we are simply talking about brain function.

Any account of the nature of human beings which concentrates on our physicality, and our kinship with other animals, will be tempted to ignore or to reinterpret our own self-consciousness. Science, indeed, typically deals with the world from an "objective" point of view. That is, it rules out any "subjective" elements in its investigations. How the world appears to me or to you, and what experience is like for each of us, is discounted in favour of an account which describes reality in terms that are the same for everyone. In one sense, this is the strength of science. What the world is like does not depend, it claims, on who is observing it, or where the science is being practised. This is indeed why the interpretation of quantum mechanics provides such difficulties for modern physicists, because it can sometimes look as if the role of consciousness cannot be eliminated from any physical account of quantum reality. Whatever the solution to that problem, its existence demonstrates the remorseless attempt of all science, and particularly physics, to remove any reference to the human mind. Yet as empiricist philosophers have long recognized, the conduct of science is firmly grounded in the possibility of human experience, through observation. If the mind is eliminated from scientific discourse, it becomes unclear what the foundations of human science can be.

Reflections like this do not worry the more aggressive of those philosophers who struggle with the problem of human consciousness, sometimes failing to recognize that their struggles are themselves the product of such consciousness, and the self-conscious use of reason. In this they have been helped by a strong tradition in philosophy since the Second World War, which has sought to reinterpret mind, without recourse to an appeal to science. The most popular way of doing this was through a consideration of our language and the way we use words. So-called "linguistic philosophy" sought to remove philosophical problems, not by confronting them, but by trying to show they were not really problems at all. We were instead merely misunderstanding our own use of language. In the words of Wittgenstein, we would be misled by "grammar". Another influential philosopher of the time, Gilbert Ryle, argued (1949 p16) that our whole understanding of the mind rested on "one big mistake". It was not a factual mistake, so much as a misunderstanding about concepts. It was a "category mistake", representing "the facts of mental life as if they belonged to one logical type or category...when they actually belong to another". We think the mind is a thing (like a house or some other substantial object) because we have a word to refer to it. Because we have a name for something, we assume there is something to be named. In fact, it was alleged, mental words are ways of describing human ways of acting. Anger, for instance, may not be a special feeling, so much as a pattern of behavior in certain situations. In fact, of course we may sometimes be able to see that a person is angry, even if the individual hotly denies it.

"Folk Psychology"

Ryle's analysis may work better for some mental "objects" than others. It is more difficult to view pain as a way of behaving, rather than as something which is felt. My first book (Trigg 1970) insisted that the distinctive felt nature of pain was the prime constituent of what it is for something to be a pain. Not all unpleasant sensations are pains, and, as I argued with examples from the scientific literature, not all pains are unpleasant. The phenomenal quality of the sensation is its defining feature, and no amount of philosophical analysis can cloak that fact. This refers to the problem of what philosophers call "qualia", the felt characteristics of experiences. Pain is a favourite philosophical example, and so is colour. What red looks like cannot be adequately accounted for in terms of wavelengths, any more than a pain can be described by reference to neural pathways and such-like. Describing an experience in terms acceptable to a scientist is to change the subject. That leaves us with the question how to relate the two descriptions. The path of the linguistic philosopher was to see the problem wholly in linguistic terns. We were talking about our use, or misuse of language, and not picking out different aspects of what makes a human being. Ryle spoke of what he termed the "official theory" of the mind, according to which we each have a privileged access to our own private stream of consciousness. He set out to repudiate this, speaking of it, as he said (1949 p15) "with deliberate abusiveness," as "the dogma of the Ghost in the Machine". This phrase has been picked up by many others and in some ways, it does characterize a traditional "dualist" view of the mind, of the kind put forward by Descartes in the seventeenth century. The phrase picks on the apparent separation of the two entities, mind and body, the one seeming to have little to do with the other. The problem of explaining their interaction then becomes especially acute. The view invited materialists to give up any idea of the ghostly entity allegedly animating our bodies, and just see the machine-like properties of human beings.

It was easy for Descartes to assume that animals did not possess minds but were mere machines, and one path for those distrustful of "ghosts" has been to see humans in the same way. However, our kinship with animals, as stressed by Darwinism, could carry a different implication, namely that some animals, at least, may have what we would recognize as mental capacities. It would be hard to deny this of the higher apes, and research tends to show that they may even have a "theory of mind", that is they can even understand how things are from another ape's point of view. Going down the evolutionary scale, it would seem extraordinary for example if birds cannot see colour – what, then, would be the evolutionary point of so much exotic plumage? The rejoinder may be that, even if they can *respond* to colour, there is no guarantee that they "see" it as we see it. This is the venerable philosophical problem of other minds. Perhaps it is idle, or even meaningless, to get involved in such speculations. Perhaps, on the other hand, there is such a thing as the viewpoint of birds, even if we do not have access to it (cf. Nagel 1974). Just because we do not have scientific access to "what it is like to be", say, a robin, does not mean that robins do not have conscious experience.

Ghosts do not seem, at least at present, to be susceptible to scientific investigation, and the response of any tough minded philosopher who believes that current science is our only method of gaining knowledge will be to dismiss anything of that ilk. Indeed Ryle's purpose in ascribing ghostly status to minds was in a sense to say that there is no such thing as mind, at least as traditionally conceived. One of his pupils who has gone on to become a prominent defender of neo-Darwinism, and a current exponent of the "new atheism", is Daniel Dennett (2006). He has replaced ordinary language philosophy with a touching faith in science. For him, what science cannot deal with cannot exist. His conclusion must therefore be the same as Ryle's, that a belief in a mind to which only a subject has access, must be ruled out. We do not properly understand mental properties, if we take them at face value. The difference between Ryle and Dennett is that while the former merely wanted to reinterpret the language he thought misled us, the latter turns to science to explain consciousness. If science turns out not to be up to the task, that is the fault of consciousness, or the way we think of it, rather than science. So Dennett is eager to deride "folk psychology" just as much as Ryle mocked the "official doctrine" of the mind. Both these phrases embody the ways we seem ordinarily, even naturally, to think. We are all dualists about the mind until we reflect on the matter.

Dennett claims (1991 p.37) that "given the way dualism wallows in mystery, accepting dualism is giving up". This is because of what he terms its "fundamentally anti-scientific stance". As we have seen, science stresses its own public character, and this must put it at odds with any claims to anything inherently subjective and private. There is, too, more than a whiff in this of old-fashioned logical positivism, of the kind popularized by A.J. Ayer – an Oxford colleague (and former pupil) of Ryle – in his own influential book, *Language, Truth and Logic* (Ayer 1946). What exists is what is accessible to scientific verification or falsification. It is the kind of "scientistic" view still championed by the likes of Richard Dawkins. What counts as "evidence" is what is acceptable to contemporary scientists following their empirical methods. "Reason" is what scientists recognize as rational through observation and deduction. Only what is judged rational in this way can be discussed in the public sphere (Trigg 2007). The problem with all this is not only that it rules out of account large tracts of human experience, concerning, say morality, aesthetics, and religion. It fails even to account for modern physics, with its reliance on theoretical and intrinsically unobservable entities. Later philosophers of science have seen the importance of theory in showing us what to look for in the world. In such a context, metaphysics, or reasoning which goes beyond what current science is capable of, may not seem so absurd. Too great a reliance on the current state of contemporary science can in fact induce complacency, and discourage us from trying to find out what we do not know. Dennett may not like mysteries, but it could be argued that it is only the recognition of one which can provide the spur to further investigation.

Dennett would no doubt retort that the character of the dualist view of mind makes it in principle beyond the scope of science. Yet science may change fundamentally in the future, so it is unclear whether that will prove wholly true. The purpose of science is to discover the nature of reality. It does not define it. If parts of reality lie beyond human science, that is because of the character of reality, and we can only avoid that conclusion by making reality, by definition, a reflection of human science. It is then a construction out of science, rather than a precondition for it. Reality becomes anthropocentric. The paradox in that situation is that reality becomes *reality for us, as experienced by us*. The stress on science becomes a stress on the scope of human experience, and reference to experience begins to refer back to the human viewpoint, which in turn breaks down into the fact of many people's subjective experiences. Any emphasis on the centrality of science cannot afford to forget that it is human science, and that, as the empiricists saw, brings us back to subjective observations and experiences. Science tries to exclude the human observer from its scope, but it must be continually stressed that all science is built on the human ability to have subjective experiences. We cannot, with Dennett, (1991 p.303) deride such experiences as "folk psychology" or forget the "official doctrine of the mind", without turning us all into machines. Yet no group of machines could ever on their own build up a body of science.

What is a "self"?

Repugnance on scientific grounds at the asserted dualism of mind and body can have major consequences, to the extent of changing our worldview. The opposite of dualism is monism, according to which there are not different kinds of realities, nor different levels. There is only one kind of stuff. Traditionally this view was known as materialism, and the opponents of Descartes often adopted an avowedly materialist outlook. There are no minds. All is matter. In more recent times, the concept of matter has proved difficult for science to pin down, as it all seems to be dissipated into energy. No longer can it be assumed that everything is made of the combination of indivisible particles called atoms. As a result, whilst it remains clear what materialism is against, it is not entirely clear what it is for. For this reason, modern materialism often goes under the names of "physicalism" or "naturalism" (Trigg 2002). These tend to limit the nature of reality not so much by describing its character as by linking it to the ability of science to gain access to it. Physicalism therefore ties itself to what makes sense in physics. It can allow for theoretical entities, as physics has to, but even so it still rules out anything that has to lie beyond our knowledge, let alone anything that has a different kind of existence from the physical. Similarly, naturalism, whilst perhaps not being so inclined to reduce everything to the terms that are acceptable to physics, also defines reality in "natural" terms, or, in other words, to what is accessible to science. In this way, ontology, or the theory of what there is, becomes merged with epistemology, the theory of what can be known. Further knowledge is still, as in positivism, ultimately linked with the methods of the natural sciences.

Naturalism is a philosophical theory, not a scientific one, and needs a philosophical justification (Trigg 2002). Why should it be decided that only a certain category of entity can exist? It is however an extremely influential position, often becoming an unargued presupposition even in philosophy. No science-based philosophy can allow for any realm beyond science, let alone "levels" of existence. Even the realm of personal experiences, including religious experiences and any putative consciousness of the divine, becomes part of nature. This is the view of Willem Drees, who points out that dualities can enter in at many points and argues that they must be opposed. He says (1996 p245):

> A duality of heaven and earth can no longer be articulated in astronomical terms, and distinctions between humans and animals and between matter and mind have become matters of degree rather than principle.

Thus humans are seen as much part of the natural world as any animal. There can be nothing that sets us apart from animals in principle, in the way that a duality of mind and body may seem to. Even if it is conceded that some animals too have mental capacities, it will be held that there can be no special substance, such as "soul" or "spirit", that can be posited to set us apart from them. Whilst this chimes in with contemporary views about evolution, it also firmly takes up a philosophical position about the kind of thing that can exist. Anything unscientific is not just beyond the reach of science. It cannot exist. Whatever else that might be said about this, it is much more than a scientific outlook. It is a bold metaphysical thesis.

We should not underestimate just how radical a change is being urged on us, compared with many traditional philosophical and theological views about human nature. It becomes much more difficult to see what is meant by saying that we are made in the "image of God", if we are just one other animal species, with no special access to anything transcendent. Indeed at risk is much more than that. What is a "self", if it is reduced to the physical workings of the brain? Dennett (1991 p.17) discards what he contemptuously terms the "Cartesian Theatre", denying the reality of our inner conscious life. Indeed he goes further, because he not only denies the independent existence of mind, but he ridicules the idea of any central controller or "central meaner" at work in the brain. There are merely parallel processes or "multiple drafts". He writes: (1991 p.434):

> There are no intrinsic 'qualia'; there is no central fount of meaning and emotion; there is no magic place where the understanding happens. In fact there is no Cartesian Theater".

The corollary, which he accepts, is that it becomes impossible to sustain the idea of a self as the subject of experience, or even as a physical unifying factor. He thus touches on one powerful factor which led some eminent scientists, such as Sir John Eccles (Popper and Eccles 1977), to champion dualism, because they felt that the metaphysical subject was an essential unifying factor in the working of the brain. It may be "unscientific" to posit such a self, but the alternative is to go against the felt experience of each of us that we are continuing persons, with the ability to combine experiences, remember them and generally compare them.

Further, whilst it is currently fashionable to see the working of the mind in wholly neural terms, this means that rationality becomes a matter of the working of the brain. It is a matter of cause and effect, not reasoning as a result of evidence, or trying to justify beliefs. Whether rational arguments in favour of physicalism or naturalism are self-referential, and succeed in undermining themselves, is a long story (Trigg 2002) but, once again, our picture of ourselves as rational, and indeed endowed with a reason that can transcend narrow physical limitations, is put in question. In that case, the seventeenth century description of reason as being "the candle of the Lord" (a view which underpinned the origins of modern science) could no longer be regarded as tenable. The causally explicable workings of the brain, provide a very different starting point from a view of reason as open to wider horizons, even those that transcend the world around us.

God and the World

This may appear just another instance of an atheist attack on the possibility of the transcendent and the supernatural. The light of reason, exemplified by science, is once more putting to flight the forces of superstition and arbitrary authority. The ideals of the eighteenth century Enlightenment are still being championed. There is some truth in this, as many of the proponents of a naturalist world-view (though not Drees) are motivated by a desire to attack theism. Whether in so doing they also succeed in undermining the possibility of human reason, and hence the validity of science, is a moot point. Certainly "post-modernists" who oppose what might be termed "Enlightenment values", are happy to accept that conclusion. It is a conclusion, however, which not only destroys the pretensions of science, but also removes any possibility of theism claiming truth, so it is not a path that should recommend itself to any who wish to assert the existence of God.

This brings us to the fact that, in discussing dualism, it must be recognized that, as Drees (1996) indicated, there are different dualities. The biggest duality of all is not that between human mind and human body. It is between God and the world, the spiritual and the physical, the Creator and Creation. Indeed it is no coincidence that, from Nietzsche on, philosophers have seen that the concepts of God and the Self are correlative. They often seem to stand and fall together. It is no coincidence that Dennett opposes them both. It is rather surprising that so many theologians nowadays are suspicious of dualism, and see it as even contrary to Christian doctrine. We are told that the doctrine of the immortality of the soul is a Greek philosophical one, and to be distinguished from the Christian belief in the resurrection of the body. The fact that this seems to fit in with modern neo-Darwinian conceptions makes it even more attractive for theologians.

Whilst it is true that a Greek notion of the soul as an impersonal, rational faculty does not fit Christian ideas of the individual person surviving death, and being under judgement, too naturalistic a notion of the human person itself carries with it great dangers. Traditional Roman Catholic ideas that somehow cremation will impede the resurrection of the body, suggests too literal a belief in the intrinsic physicality of the identifiable person or "body". A glorified, transformed body, presumably in a different form of existence, should not too readily be identified with our finite condition here and now. St Thomas Aquinas got himself into great difficulties over this (Trigg 1999 p.40).

Whatever is to be understood as the resurrection of the body, the fact remains that seeing it in physical terms does not meet an underlying metaphysical difficulty for anyone who wants to see this in terms of theistic belief. Talk of the human person as a "psychosomatic unity" may seem to make Christian ideas of the person more scientifically respectable, by, in effect, omitting reference to anything beyond science's reach. We still, though, have the issue of the status of God. If the notion of an immaterial self is to be regarded as somehow illegitimate, does not the same reasoning apply to the concept of God? In fact, there are many philosophers of religion and theologians who accept this conclusion, and try, in radical fashion, to reinterpret the idea of God as a metaphysical entity. One ploy is simply to see it in terms of human reactions, perhaps of awe and wonder, to the world we find ourselves in. Alternatively, pantheism has always tried to identify the divine with the workings of the natural world. More recently "panentheism" has also adopted a more naturalistic stance, connecting God with the processes of the world. There does however, in that case, seem to be some metaphysical remainder, distinguishing it from pantheism, so some ultimate duality between God and world still remains.

Any attempt to reinterpret ideas of God to avoid the challenge of metaphysics, and to remain within a physicalist world-view, has to be resisted if new internal contradictions for theology are not to be produced. For example, the ultimate separation between God and the world is an essential part of dealing with the problem of evil. Any identification of God with the processes of the world removes all distinction between how things are and how they ought to be. Without a distinction between a finite and transitory world and an eternal, spiritual one, theism becomes reduced to wishing, in a hopeless manner, that the world was very different from the way it is. The flawed nature of the world we inhabit is all too apparent. Both natural and moral evil become even more intractable problems, if we conclude that what is physical is all there is, and that there is no higher, or different realm, against which the deficiencies of our existence can be judged.

Some see dangers in this approach. "Does not this mean", they would say, "that we are led to devalue the world we inhabit, and to hold in contempt the material?" Those who figuratively set their eyes on the heavens are, in the old joke, no earthly good. Whereas, it will be pointed out, Christians should not take the existence of another realm as an excuse for not caring about this one. This, it will be said, is the fault of Platonism, devaluing and despising the physical world, because all attention is given to some world of absolute perfection. This theological objection does have some force. Any dualist scheme which places excessive weight on one form of entity at the expense of the other is not in fact genuinely dualist. It is sliding into a form of monism, in this case placing the spiritual far above the material. If there are two kinds of reality, ignoring or even devaluing one is in effect beginning to deny its genuine reality. This is in fact what Plato himself did, at least in some of his dialogues, by restricting terms such as "knowledge" and "existence" to the world of Forms. We could only have beliefs about the material world as the world of change.

This, though, is not the Christian view. There are echoes of Plato in the New Testament, as when St Paul famously distinguishes in 1 Corinthians 13 between now "seeing through a glass darkly", and then seeing "face to face". That does not mean that we cease bothering with the here and now. The whole point of the Christian understanding of Incarnation is that the Word became flesh. Two levels of reality were joined. Our corporeal existence was sanctified, and shown to be of crucial importance to God. The two worlds were bridged.

Theism and Dualism

An abhorrence of "ghosts" and spirits may be good scientific methodology. Science has to look for causal explanations which are accessible to it, not blame events on non-material entities which cannot be verified. In this sense, bringing in the non-material is "giving up" for a scientist. It is like blaming events on the fairies at the bottom of the garden. At the same time, science should be concerned not just with its own capabilities, but with the nature of reality. There is no *a priori* reason that can be produced, without philosophical argument, to demonstrate that science does not have its own intrinsic limitations. There may be other paths to knowledge. Not all reality may be accessible to human scientific method. Metaphysics is unavoidable, even if the metaphysical position one adopts is that only the physical can exist! A theist must recognize that, unless the concept of God is being changed into something unrecognizable, God is a Spirit, and is not a part of the material world, or to be identified with it in any way. Thus theism is, of its nature, on a collision course with physicalism, or a metaphysical naturalism. Indeed when physicalists and naturalists attack any notion of the supernatural, it is usually God who is prime target.

Whilst some radical thinkers may feel that this means that we must give up any traditional understanding of God as non-material, a moment's reflection will make it clear that this must be a step too far. At that point, theism has by definition been abandoned in favour of some form of romantic, or mystical, humanism. The idea of an infinite God, as separate from the finite, imperfect world that has evolved, is a crucial one. Our world, and the universe of which it is a part, had a beginning and will have an end. God, to be God, cannot come into being or go out of being. Some may sneer that this is the "God of the philosophers", but there would be little point in the idea of revelation if God was identifiable with a reality wholly accessible to us. The need for revelation only occurs because two realms of reality have to be put in contact with each other.

If, too, we believe that God truly is our Creator, we may feel that He will not have left us stranded in an alien world with no means ourselves of understanding anything of His reality. The idea, already mentioned of reason as the candle of the Lord is derived from some such notion. Current research in the cognitive science of religion, which itself draws heavily on evolutionary psychology, may suggest that there is some natural sense within us of something divine, transcending our own form of existence. Certainly religion appears to be a universal phenomenon in human nature. Some of this remains speculative, but what is crystal clear is that our understanding of ourselves and of our nature, is indissolubly linked with the question of a belief in a transcendent God. If we do not believe in God, this affects our understanding of ourselves. Similarly, if we are driven to a naturalist understanding of what it is to be human, this is bound to affect our willingness to believe in anything supernatural, anything apart from ourselves and our physical world.

Do we think of humans as only evolved animals, wholly explicable in physical terms? Do we, on the other hand, see ourselves as having an eternal destiny, and this life as part of a wider whole? How we see ourselves is inseparable from whether we think there is another, spiritual, realm, perhaps permeating, but certainly going beyond this one. Any world-view, based on science, and limited by its capabilities, will recoil with horror from such speculations. Yet our unreflective understanding of ourselves is dualist. *The Self and Its Brain* was the title of the book advocating dualism written thirty years ago (1977) by the neurobiologist, Sir John Eccles, and the philosopher Sir Karl Popper, and it exemplifies the issue facing us. Is the "self" different from the brain? If it is, there are, of course, considerable difficulties in associating the two elements of our nature, the physical and the non-physical. It may not be inappropriate, without disparagement, to call it a "mystery". If, though, the two are to be identified, or if the idea of a self is to be denied, we not only go against our "natural" understanding and the seemingly clear facts about self-consciousness. We land ourselves in a major theological and philosophical conundrum, if we still wish to espouse theism. There are definite analogies between the relationship of the self and the brain, and that of God and the created cosmos. Indeed, the philosopher of religion, Charles Taliaferro (1994), draws an explicit parallel between the view that "persons are not identical with their bodies" and that "God is not identical with the cosmos". Even issues about their interaction, and possible divine intervention in the physical world, are themselves shrouded in mystery. That will be enough for many to suppose there can be no such relationship, though that depends on the assumption that reality must conform to what we humans, here and now, are able to understand. It is a view impatient of any admission of human fallibility, and of intrinsic human limitations. It is, incidentally, also a position that does not allows for any unexpected developments in science which radically change our current understanding.

In recent years, much has been made of the concept of emergence in connection with mind. Mental properties are seen as "emerging" from the physical, and this may seem to fit in well with an evolutionary account of mind. It may indeed provide an account of a pathway of development which may still be seen leaving its traces in various animal species. This however may provide a story of how mind has evolved, but it does not tell us what mind *is*, or deal with its metaphysical status. It does not tell us whether our mental life is real, nor does it deal with the relationship of the mental and the physical. Is what has emerged totally dependent on its physical substratum, or, having emerged, can it exercise a causal influence even on our bodies (including our brains)? In other words can there be "top-down" as well as "bottom-up" causation? If there can be, and so-called psychosomatic illness might be one just example, we are back with a distinctive dualism of two realms, or categories of reality, interacting with each other. Those who toy with concepts such as "non-reductive physicalism" also want to have it both ways. They want the advantages of monism, so that they do not have to defend what looks like an "unscientific" view. At the same time, they are reluctant to deny the facts of our mental life, which seem so real to each of us. The point is that if the mind cannot be reduced to something else, and is not totally impotent, it must have some form of an independent reality, which allows it to interact with our body.

Physicalism, and naturalism, have powerful friends. Theists, however, must not be seduced into thinking that they can somehow subscribe to a physicalist account of the person, and still retain an unrevised theism. If we are wholly material beings, it makes little sense for us to posit some form of non-physical ultimate reality. There is little point, from a metaphysical point of view, in being monists about the mind and body, but dualists about God and the world. Both are liable to stand and fall together. We may want to wield Ockham's razor, and cut out superfluous entities, but that will apply to God as well as the self. The issues of the existence of the mind and that of God are, however, not just parallel ones. They are intimately connected. If this is a world created by God, then theism can itself provide grounds for seeing why humans are as they are. As Taliaferro says (1994, p.84):

> A theistic outlook will provide a fuller model of explanation in which the natural emergence of the mental from the physical, and indeed the very constitution and powers of the physical world itself, is seen as stemming, from a deeper, underlying cause.

We do not just conclude that there are minds, and independently think that there is a God. Our minds enable us to search for the divine, and, the Creator can be seen as deliberately making creatures in His image, who can respond to Him. If God is Spirit, but we are wholly material, it becomes far from clear how, or why, we can respond to Him.

A dualist view of the human person does not deny our evolutionary origins, or our physical nature. It is not an "idealist" belief, saying that minds are the only reality. On the other hand, it refuses to go with the fashion of the day and say that we only physical. It claims that there are two orders of reality, each important. The success of science inclines many to look only to science as a source of our understanding of human nature. Dualism is not the preponderant view amongst contemporary scientists or philosophers or even, more surprisingly, of many theologians. Rather than thus blindly following the spirit of the age, it may perhaps be important for us to see that this puts any traditional understanding of God in jeopardy. Physicalism in any form cannot be combined with theism, without radically affecting the content of religious belief. The fact that it challenges our own everyday experience of ourselves should also perhaps give us pause for thought.

References

Ayer, A.J. 1946. *Language, Truth and Logic* (2nd ed.), London: Victor Gollancz.
Dennett, D. 1991. *Consciousness Explained*, Boston: Little Brown.
—. 2006. *Breaking the Spell, Religion as a Natural Phenomenon,* New York Viking.
Drees, W. B. 1996. *Religion, Science and Naturalism*, Cambridge: University Press.
Nagel, T. 1974. "What is it like to be a bat?" *Philosophical Review*, 83, 435-450.
Ryle, G. 1949. *The Concept of Mind*, London: Hutchinson.
Popper K, & Eccles, J.C. 1977. *The Self and Its Brain*, Berlin: Springer.
Taliaferro, C, 1994. *Consciousness and the Mind of God*, Cambridge: University Press.
Trigg, R. 1970. *Pain and Emotion*, Oxford: University Press.
—. 1999. *Ideas of Human Nature* (2nd ed), Oxford: Blackwell.
—. 2002. *Philosophy Matters*, Oxford: Blackwell.
—. 2007. *Religion in Public Life*, Oxford: University Press.

CHAPTER NINE

THE EMERGENT THREEFOLD SELF: A RESPONSE TO ROGER TRIGG

ANNE L.C. RUNEHOV

Dr Anne Runehov is an Associate Professor of Systematic Theology in the University of Copenhagen. Belgian by upbringing, she married into Scandinavia, and turned to advanced study only after raising her family. She took an MA in philosophy (especially of mind) from Stockholm and a ThD in philosophy of religion from Uppsala, before moving to Copenhagen. Her first full-length book was, "Sacred or Neural? The potential of neuroscience to explain religious experience" (2007), but she is now writing prolifically; among her major tasks is that of philosophical section editor for the forthcoming "Encyclopaedia of Sciences and Religions".

In this tightly-argued response to Roger Trigg, Anne Runehov agrees that conceptions of the self have inescapable implications for conceptions of the God-world relation too; but she proposes that the strengths of both monist and dualist positions can be united in one that is actually trinitarian in its accounts both of personhood and of overall reality. At each level her concept is not of three separate stuffs but – as in traditional Christian concepts of God – three aspects of one substance.

Introduction

Are we ghosts or machines? Professor Trigg's title reminds me of my own book in which I asked the question whether religious experiences are Sacred or Neural. My answer was that they are both sacred and neural (Runehov 2007). Analogically, could the answer to Roger Trigg's question perhaps be that we are both ghosts and machines? According to Trigg "there are two orders of reality", a physical/material reality, which science can investigate, and a non-physical/non-material reality, which science cannot by itself either study or explain in terms of the physical/material; neither can science simply ignore it. I agree with there being (at least) two orders of reality; however, I disagree with the limits imposed on science. Hence, my response, though accepting many of the ideas put forward by Trigg, will argue for non-dualistic comprehensions of both the self and the God-world relationship. My viewpoint is based on contemporary neuroscientific research about the self, Trigg's two orders of reality being taken into account. The suggested model is monistic in nature, i.e. as Trigg argues, there is "only one kind of stuff"; however, this stuff will be shown to be threefold. In the shorthand of symbolic logic, the following explanatory models are suggested (Runehov 2008b):

1. $ES((ONS \leftrightarrow (SNS \cap STS)); STS > (ONS \cap SNS)$

where ES stands for one Emergent Self comprising an Objective Neural Self (ONS), a Subjective Neural Self (SNS) and a Subjective Transcendent Self (STS). There is mutual causation ($\leftrightarrow$) between the objective and subjective neural selves, i.e. between ONS and SNS plus its intersection ($\cap$) with STS. Furthermore, the subjective transcendent self is bigger (>) than both the objective and subjective neural selves. This ES, being the sum of the different elements of the self (ONS, SNS and STS) will, from a Judaeo-Christian theological perspective, be understood as the *imago Dei*:

2. $EU(G \leftrightarrow (NR \leftrightarrow ES)); ES > NR \cap G > (NR \cap ES)$

where EU stands for one Emergent Universe comprising God (G), Natural Reality (NR; world) and all Emergent Selves (ES). There is mutual causation between God and the world, between the world and Emergent Selves (the ES being part of it) and between God and Emergent Selves within the world. Furthermore, the Emergent Selves are bigger than the world and God is bigger than both the world and Emergent Selves.

Is there something special about the human self?

As Roger Trigg also notes, Daniel Dennett would answer this question in the negative; there is nothing more to the human self than physical and material reality. We behave and experience as we do simply because natural selection ("Mother Nature") brought our brains and nervous systems into being in this way, though not deliberately (Dennett 1993). Dennett's answer is an ontological reductionist one, and fails to consider the mutual causal connection that seems to occur between the neural and the mental. Many contemporary neuroscientists would not agree with Dennett's view. Neuroscientists, when studying the self, for example, also take the subjective descriptions of the patients or participants into account. It is true, as Trigg argues, that they cannot tell us *what* the self is, but they can tell us *why* neurologically we behave and experience as we do; and I diverge from Trigg in viewing this subjective behaviour and experiencing as threefold, mirroring a threefold self. The self is a very complex issue. It refers to our experiences, our relations with and understanding of, ourselves and others and, for a lot of people, their understanding also of God; our feelings and emotions, our attitudes and behaviour, our thinking, dreaming, and so forth. Therefore we expect the correlating neural activity to be complex as well, if it is to sustain the self in all its expressions. Indeed, neuroscientists explain the self neurologically in terms of complexity, a *nested hierarchy,* which implies that the elements comprising the lower levels of the hierarchy are physically nested within higher levels to create increasingly complex wholes (Feinberg & Keenan 2005). However, when we are in the process of thinking, for example, we would expect specific neural activity to correlate with that thinking, even though it would be embedded in the nested neural hierarchy. We expect this correlating neural activity to be specific in the sense that it would show an unmistakably altered activity during a process of thinking compared to when there is no such process. Indeed, neuroscientists agree that for every specific subjective self experience there seems to be a specific neural correlation within the frontal, parietal and temporal regions of the brain (Feinberg & Keenan, 2005, Decety *et al.* 2003, Johnson *et a*l. 2002, Fossati *et al.* 2003).

The relationship between the neural and subjective selves

Because neuroscientists explain the self in terms of a nested instead of a non-nested complex hierarchy, they dismiss a dualistic view of the relationship between the neural self and the subjective self. A *non-nested* hierarchy would imply there is a hierarchy of independent entities: a nested one implies no such independence. Furthermore, whether dualists defend strong or essential dualism, transcendental dualism or implicit dualism, they all struggle with the problem of filling the gap between the physical and the mental. Still, it might not be wise to dismiss the dualist comprehension of the self. As Roger Trigg rightly points out "our unreflective understanding of ourselves is dualist". Hence, if we want a holistic explanation of how the brain and the mental processes are related, covering both neuroscientific findings (dismissing dualism) and subjective emotional descriptions, we need to find an explanation that takes the dualistic understanding of the self into account *without giving into dualism.* I also agree with Trigg that materialism and physicalism are not fruitful options. These views are unable to describe what it is like to have such-and such an experience, which means that they deprive human beings of their experiential reality. That leaves us with non-reductionist models of the relationship between the brain and the mind. Non-reductionists argue that there are states, processes and events over and above bodily states, processes and events. Often non-reductionists rely on a combination of the Token Identity Theory (Lockwood 1989), and the Principle of Supervenience (Lockwood 1989, Murphy 1998). According to Token Identity theorists, particular brain tokens *correlate* with particular mental tokens. Taking Token Identity together with the Principle of Supervenience, such theorists can argue that there is a difference at the physical-state level for every difference at the mental state level, on which the difference between the relevant mental states in some sense depends (Lockwood 1989 p.21). However, even though mental states depend on physical states, they are not reducible to them. Because of the irreducibility, Philip Clayton and Jaegwon Kim call this type of supervenience, *Weak Supervenience* (Clayton 2004 pp.124–125). This *dependency relationship between mental states and the brain* is what neuroscientific research on the self reveals. However, as Philip Clayton argues, non-reductionism has its problems. Indeed, non-reductionists need to explain the "over and above the physical or material", without falling into dualism, physicalism or materialism. To put it in Trigg's words, they do "not tell us whether our mental life is real".

However, Michael Lockwood argues that mental states and events are real because they follow the *same temporal and spatial order* as physical states and events. Indeed, supported by Einstein's Special Relativity Theory, he assigns independent locality to mental states, i.e. mental states must be in space given that they are in time. Lockwood's move is a fruitful one. What is not clear is whether this relation between the mental and physical also implies mutual causation. I hence suggest that the Principle of Strong Emergence (PSE) be added to the explanatory model, because:

> ... if life and mind are genuinely emergent, then living things and mental phenomena must play some sort of causal role; they must exercise causal powers of their own. (Clayton 2004 p.30).

Indeed, our brains change continually in relation to what we learn, read, and experience. My brain has changed since this morning; yes, since I have written this sentence. Furthermore, Newberg and d'Aquili discovered that the meditators in their study had a significantly different "thalamic laterality index" at baseline compared with the control subjects. In other words, there was a difference in the neural propensity in the thalamic brain function *before* the acts of meditation took place. Thus this difference is *constant*. The differences in the thalamus led them to assume that the meditators might have undergone changes in their brain due to acts of meditation (Newberg *et al*. 2001 pp.117–122). It seems that the different thalamic laterality index is due to the emergent properties initiated by mental activity.

The threefold self

There is as yet no consensus for what or who the self and its function might be and I do not claim that I will solve the problem. However, let me suggest a notion that is in accordance with neuroscience, philosophy and theology. Thus far we postulated an emergent self (ES) comprising an objective neural self (ONS: the neural correlations of all the possible expressions of the subjective self) and a subjective neural self (SNS): if objective neural functions are damaged, so are the corresponding subjective expressions. However, neuroscientific results such as those of Newberg & d'Aquili seem to point to a third parameter; let us call it *the subjective transcendent self* (STS). I suggest that it is this STS to which Trigg refers as the "dualistic understanding of the self".

The subjective transcendent self has been explained in different ways. For instance, neurologist Harald Walach names it "consciousness", meaning by this the "personal awareness of being myself and knowing this fact. ... [T]he 'internal' side of our being" (Walach 2007 p.216). Lockwood also refers to it in terms of consciousness and explains it as a searchlight:

> What we see are the objects that the searchlight illuminates for us. We do not see the searchlight. Nor do we see the light: merely what the light reveals" (Lockwood 1989 p.169).

Hume could not discover impressions about himself (Hume 1978 [1739-40] p.272). Husserl speaks of an inner consciousness that is always present. Heidegger introduced the term *Dasein,* which is the "entity which in each case I myself am" (Seigel 2005 p.571). Norton Nelkin distinguishes between the *being in-control* and the *being not-in-control*, which to him is the basis for the distinction between *me* and *not-me* or *self* and *not-self*. Furthermore, according to him, there is an *underlying essential subjectivity*, an introspective capacity that makes one aware of this distinction in one's own experience. Indeed, human beings realize when their bodies and their minds are out of control. For instance, Reverend Bob Davies, writing in his diary about his Alzheimer's disease, clearly demonstrated this. He wrote, "When the darkness and emptiness fills my mind, it is totally terrifying. I cannot think my way out of it (Hopkins 1997 p.82).

Hans Runehov, a computer scientist, argues that in the case of Reverend Davies, the brain seems to have a *read-only capacity*. Whereas, in normal circumstances, there is a read-write collaboration between the neural and subjective selves SNS and STS, under specific circumstances the STS seems to have a *restricted read-write* capacity and under the worst circumstances (as in Reverend Davies' case) the capacity is merely *read-only* – it is not in control of the self and not-self but remains observant. Split-brain patients may, in their turn, illustrate a restricted write capacity. One of Kathleen Baynes' patients could write only out of the hemisphere not associated with language. According to Baynes, the ability to write seems to "... stand alone and does not need to be part of our inherited spoken language system"(Gazzaniga 2008). Of course, due to the nested neural hierarchy correlating with the nested subjective hierarchy, one expects there to be some neural support for the writing action in some way. Similarly it can be argued that in some way there is neural support in order to make the subjective self able to remain observant, even during the very last stage of Alzheimer's, when almost every function of the brain has decayed. Exactly which neural function(s) are the supporting factors so far remains undiscovered, but it seems that this neural support does not derive from the neurofunctions expected to correlate with such behaviour. Hence, the mental self seems to have a restricted authority, i.e. it is able to *command* the neural circuitry in a restricted manner.

We have postulated a (strong) emergent threefold self consisting of a neural self, a subjective neural self and a subjective transcendent self. The relationship between the three elements is as follows: ES((ONS ↔ (SNS ∩ STS); STS > (ONS ∩ SNS). The function of the ONS is to neurologically sustain the subjective selves. The function of the SNS then is to express the neural self. Finally, the task of the STS is to be the essential observing subjective self, to be the self that always was and always is itself, irreducible to neither the neural self nor the subjective neural self, i.e. *dasein* and being in optimal control. By way of mutual causation, the three elements of the self cause the emergent process of the whole self (ES).

The *imago Dei*.

The question to be answered now is how to understand the emergent threefold self theologically. As Wentzel van Huyssteen argues:

> For Christian theology one of the most crucial questions today should be whether there is a way in which we may rediscover the canonical function and orienting power of a concept like the *imago Dei* without retreating into metaphysical abstractions [and thus] facilitate interdisciplinary reflection? (Van Huyssteen 2006)

In short, being *imago Dei* means possessing qualities, relationships and functions similar (but not equal) to God's and using them similarly (but not equally) to the way that God would do. Postulating the *imago Dei* view means postulating that the purpose of the emergent threefold self is to be all that. The *imago Dei* comprises both the mental (including the spiritual) and the neural and is in accordance with neuroscientific and the philosophical as well as theological understandings of the self.

Panentheism

If we want to understand a God-relationship with the world in an interdisciplinary manner that also includes theology and furthermore, that corresponds to the proposed model of the emergent threefold self as *imago Dei*, I belief panentheism needs to be reconsidered as a potential explanatory model of such a relationship. The advantage is that we keep the Christian values Roger Trigg refers to: "Christianity needs a separation of the Creator and the created world"; but we also have the Judaeo-Christian concept of God being both immanent and transcendent in relation to the world. To be a fruitful model of panentheism, at least the following elements are desirable.

1. There is only one universe (U) comprising both ultimate (UR) and natural reality (NR).
2. God created the world from within Godself and interacts with it from within the world through the nested hierarchy of the world by way multiple causations.
3. God is immanent and transcendent in relation to the world and human beings
4. There is emergence; hence, the one universe U becomes EU.

5. Because of the threefold view of the self and because human beings are understood as *imago Dei,* the panentheistic view includes this doctrine.
6. From this it follows that the world becomes the mediating interlacement between God and human beings.

The first, second, third and fourth elements are covered by various models, notably Arthur Peacocke's panentheistic one which implies that there is only one closed universe that includes both the natural and an Ultimate world (Peacocke 2004). Hence, God is immanent and transcendent in relation to the world and human beings. Peacocke's and Clayton's panentheistic models also emphasize there being a process of emergence (Clayton & Peacocke, 2004 pp. 87–88). God is relationally in the world but is not the sum of the world and, simultaneously, human beings are relationally in God. This is in analogy with the mutual relationship between the threefold self and the brain. The threefold self is relationally in the brain but is not the sum of the neurons and, simultaneously, the neurons are relationally in the self through different neural correlations.

The justification for including elements 5 and 6 in the panentheistic model derives from its early formulation by Nicolas of Cusa (1401-64). His panentheistic model already includes the concept of the *imago Dei* (Runehov 2008a). Nicolas also understands the created world as the Creator becoming visible, as God's similitude. God, the world and human beings are not separate realities; they are parts of one reality, with a structural tension between them (Sandbeck 2007). However, even though human beings are God's similitudes, just as the natural world is, they are also God's images. By "similitude" is meant that the world is God presented as a multitude of real things but the real things (as a whole or in part) are at the same time representations of God. Human beings are also God's images, *Deus secundus*. Human beings in turn create rational things. The world now becomes the mediating interlacement between God and *imago Dei*. Reality becomes a dynamic process: God, the world and human beings are continuously changing due to an emergent causal trinitarian process. God created the world from within Godself and interacts with it from within the world through the nested hierarchy of that world. $EU(G \leftrightarrow NR \leftrightarrow ES); G > (NR \cap ES)$.

Conclusion

Instead of arguing for a dualistic understanding of the self and the relationship between God and the world, which I see as problematic from an interdisciplinary point of view, I propose a trinitarian view, which I believe to be more fruitful and more in agreement with the Christian faith. Because I understand the human self in terms of one closed emergent threefold self, there are no gaps between the physical and the mental (spiritual) requiring explanation. The emergent threefold or trinitarian self includes; *the neural self* (ONS); *the subjective neural self* (SNS) and *the subjective transcendent self* (STS). There is at least triple causation between the ONS, SNS and STS, giving rise to new and irreducible properties by way of a process of emergence. The mental properties are real in the same sense as the non-mental are, i.e. they follow the same temporal and spatial order. From a theological point of view, the emergent threefold self represents the *imago Dei*, possessing qualities, relationships and functions similar (but not equal) to God's. The purpose of human life is to use these abilities similarly (but not equally) to the way that God is understood to do.

Instead of defending traditional Theism, I have proposed a Panentheistic model in order to explain the God-world-human relationship. This view is also trinitarian in nature. Again, because I understand the universe as one closed emergent system including both the Ultimate and natural realities, there are no gaps between the physical and the mental (spiritual) requiring explanation. There is at least triple causation between God, the world and human beings, giving rise to new and irreducible properties by way of a process of emergence. From a Christian point of view one might argue that God's incarnation in Jesus Christ is the perfect emergent result of the causal affection between God and the world. To put it in Roger Trigg's words, "two levels of reality were joined, the two worlds were bridged". Unfortunately, I do not have space here to discuss the problem of evil within a panentheistic view, but it is possible so to do (Runehov 2008b).

We are both ghosts *and* machines, i.e. one threefold self consisting of a machine to sustain the ghost, a part of the ghost that correlates one-to-one with the machine and a part of the ghost that transcends the two.

References

Clayton, P. 2004. *Mind & Emergence, From Quantum to Consciousness*, Oxford: University Press.
Clayton, P. & Peacocke, A. (eds.) 2004. *In Whom We Live and Move and Have Our Being*, Grand Rapids: Eerdmans.
d'Aquili, E.G. & Newberg, A.B. 1999. *The Mystical Mind: Probing the Biology of Religious Experience*, Minneapolis: Fortress Press.

Decety, J. & Sommerville, J.A. 2003. Shared representations between self and other: A social cognitive neuroscience view, *Trends in Cognitive Science*, 7, 527-533.

Dennett, D. 1993. *Consciousness Explained*, London: Penguin Books.

Feinberg, T.E. & Keenan, J.P 2005. Where in the brain is the self? *Consciousness and Cognition*, 14, 661-678.

Fossati, P., Hevenor, S.J., Graham, S.J *et al.*, 2003. In Search of the Emotional Self: An fMRI Study Using Positive and Negative Emotional Words, *American Journal of Psychiatry*, 160, 1938-1945.

Gazzaniga, M.S. (01/02/2008) "The Split Brain Revisited. Groundbreaking work that began more than a quarter of a century ago has led to ongoing insights about brain organization and consciousness", http://cwx.prenhall.com/bookbind/pubbooks/morris4/medialib/readings/split.html.

Hopkins, D. 1997. "Failing Brain, Faithful Community", in *God Never Forgets*, ed. D.K. McKim, Louisville: Westminster John Knox Press.

Hume, D. 1978 [1739-40]. *A Treatise of Human Nature*, ed. L.A. Selby-Bigge, Oxford: Clarendon Press.

Johnson S., Baxter, L.C., Wilder, L.S.*et al.*, 2002. Neural Correlates of Self-Reflection, *Brain*, 125, 1808-1814.

Lockwood, M. 1989. *The Mind, the Brain and the Quantum*. Oxford: Blackwell.

Murphy, N. 1998. Supervenience and the Nonreducibility of Ethics to Biology, *Evolutionary and Molecular Biology: Scientific Perspectives on Divine Action*, ed. R.J. Russell, W.R.Stoeger & Ayala, F.J., Vatican City State / Berkeley: Vatican Observatory / Centre for Theology and the Natural Sciences, 463-489.

Nelkin, N. 2001. Subjectivity, *A Companion to the Philosophy of Mind*, ed. S. Guttenplan, Oxford: Blackwell.

Newberg, A.B., Alavi, A., Baime, M., *et al.* 2001. The Measurement of Regional Cerebral Blood Flow during the Complex Cognitive Task of Meditation: a Preliminary SPECT Study, *Psychiatry Research (Neuroimaging Section)*, 106, 113-122.

Peacocke, A. 2004. Articulating God's Presence in and to the World Unveiled by the Sciences, in *In Whom We Live and Move and Have Our Being*, ed. P. Clayton & A. Peacocke, Grand Rapids: Eerdmans, 137-154.

Runehov, A.L.C. 2007. *Sacred or Neural? The Potential of Neuroscience to Explain Religious Experience*, Göttingen: Vandenhoeck & Ruprecht.

—. 2008a. God, Linné and Dawkins. *Uppsala University* (in press).

—. 2008b. Neuroscientific Research on the Self: A Case for Panentheism, Metanexus online publications:
http://www.metanexus.net/conference2008/articles/Default.aspx?id=10513
Sanbeck, L. (ed.) 2007. *Mig og evigheden: Johannes Sløks religionsfilosofi*, Copenhagen: Forlaget Anis.
Seigel, J. 2005. *The Idea of the Self*, Cambridge: University Press.
Van Huyssteen, W. 2006. *Alone in the World? Human Uniqueness in Science and Theology*, Grand Rapids: Eerdmans.
Walach, H. 2007. Mind – Body – Spirituality, *Mind & Matter*, 5, 215-240.

CHAPTER TEN

UNFOLDING CONVERSATION: A THEOLOGICAL REFLECTION ON THE EVOLUTION OF THE BRAIN/MIND[1] IN *HOMO SAPIENS*

JEREMY LAW

Revd Dr Jeremy Law is Dean of Chapel at Canterbury Christ Church University. He originally studied geology, at both undergraduate and postgraduate levels, before going on to take a theology degree as part of his training for the Church of England ministry. After ordination, he obtained a doctorate from Oxford University in Systematic Theology. He then spent nine years as Lazenby Chaplain and Lecturer in Theology at Exeter University. He moved to Canterbury in 2003.

In this chapter, an extended version of his plenary lecture at the 2007 conference, Dr Law melds the scientific and theological strands of his own outlook to construct impressively full and rounded accounts of humanity and of Christ that are simultaneously paleo-zoological and theological.

The Question of Meaning

The Ardeche Gorge of south-eastern France hides a remarkable treasure. Accessible only via a labyrinthine passage and far from the reach of daylight, lies a set of exquisite cave paintings. Here on the walls of the *Grotte Chauvet* can be found a profusion of animals: they include lions, deer, rhinoceroses, panthers and buffalos (McKie 2000). Each is captured

---

[1] The formulation brain/mind is used so as to distinguish the physical and mental aspects of cognition while holding that the latter is irreducibly related to the former.

with a skill that results in lifeless paint being transformed into a living representation ready to leap from the inanimate rock. Dating to roughly 32,000 years ago, and thus fully 15,000 years older than the more famous cave art of Lascaux, these images constitute the earliest concrete evidence in Europe of the emergence of a mind like our own – a *Homo sapiens* mind (Lewin 1998 p.470). These images are the product of minds that understand symbolism and have harnessed that power in the search for meaning.

Behind the emergence of minds like our own, is the vast hinterland of the evolution of life by natural selection. It is estimated that the lineage of development that led ultimately to the arrival of *Homo sapiens* diverged from our closest cousins, the chimpanzees, between 6 and 5 million years ago (Mithen 2005 p.107). This divergence, however, was as yesterday compared to the almost 4 billion year process of life itself. You who read, and I who write, are alive today because we are part of a continuous, unbroken stream that reaches right back to life's ultimate origin. Yet, according to Neo-Darwinian orthodoxy, there is no discernable scientific reason why we, *Homo sapiens*, had to evolve. Hence there is no scientifically discernable purpose to our existence.[2] We are the products of a thoroughly contingent process with no necessary destination (Forrest 2000). For Stephen Jay Gould, a trenchant defender of this view, "*Homo sapiens* is not the foreordained product of a ladder that was reaching toward our exalted estate from the start" (1980 p.62). In fact, for him, we are a mere epiphenomenon of a process of the diversification of life that had nowhere to go from its simplest beginnings but in a more complex direction (1996 p.172). Moreover:

> If we could replay the game of life again and again...the region of [life's] greatest complexity would be wildly and unpredictably different in each rendition – and the vast majority of replays would never produce (on the finite scale of a planet's lifetime) a creature with self-consciousness. Humans are here by the luck of the draw. (*Ibid.* p.175).

I wish to argue, therefore, that the *distinctive* theological challenge provided by evolution through natural selection, of life in general and of *Homo sapiens* in particular, is to be found in meaning. It is difficult to reconcile human existence as sheer fluke with the notion of the world as the intended creation of God. An evolutionary perspective also, of course, casts a particular light on the issues of suffering, death and wastefulness that constitute the classic areas of theodicy (the attempt to justify the ways

---

[2] On this general theme, cf. Fraser Watts' comments, Chapter Four. Ed.

of God in the light of the nature of the world[3]. However, these issues also pertain within a creationist purview. Even the extinction of species is observable as a present phenomenon. Moreover, the process of evolution provides a theological advantage in this task of theodicy over the essentially static view of the past, for it affords the inevitable death of individuals a new significance. Death has been the engine of development of protean forms of life. It is, then, the question of meaning with which we shall be concerned, as the primary challenge to a theological interpretation of human existence.

One option, taken up by Barbara Forrest (2000), is what might be termed the existentialist approach. Here, she is influenced by Daniel Dennett's *Consciousness Explained* (1993). Given that there is no intrinsic meaning or purpose to human existence, it is our task, as those who are characterised by an ability to pose questions of meaning, to create that meaning for ourselves. This approach, however, appears to both affirm and simultaneously undercut our defining characteristic, for it could always be said of any asserted meaning: "that's simply what you have created!" In this way, meaning is rendered arbitrary, as a matter only of persuasive rhetoric and so an exercise of power. If meaning matters, it is, I suggest, because we are *makers* of meaning as a corollary of our nature as *seekers* of meaning. We are not simply interested in "meaning for meaning's sake". Rather we find ourselves at the heart of a struggle to discern accounts of meaning that – to use a currently much despised term – lay claim to being truer than available alternatives. And we do so even if this process is intrinsically incapable of final resolution.[4]

Theological Reflection and Meaning

There can be no meaning without interpretation. It is not until our observations are placed within a "horizon of interpretation" (Gadamer 1975) – a pre-existing cluster of presuppositions about significance – that they can be rendered meaningful. This hermeneutic process takes place so commonly and so pervasively that it is often unobserved. To bring this process to consciousness, think, for example, about the interpretive strategy that takes place when one meets another for the first time. One cannot but start to form an impression of the person based on what is observed, while this is judged by one's whole previous history of inter-personal encounter (the horizon of interpretation in this case). Commonly,

---

[3] See for example, the work of Christopher Southgate (2008).
[4] To explore this issue further see Richard Bailey (2001).

however, this initial interpretation will shift, by varying amounts, as one becomes better acquainted. This process engenders an unending spiral of interpretation as person and horizon of interpretation shift in relation to each other.

To return to the matter at hand, it follows that every attempt to lay bare the process of evolution takes place within such a cluster of presuppositions about significance. More grandly, every attempt to understand the process of evolution, whether in scientific terms or not, carries an implied metaphysic (cf. Conway Morris 1998 p.14).

Theological reflection on the evolution of the brain/mind in *Homo sapiens* will seek to create meaning by reference to God – that is its identifying distinctiveness. For myself, a Christian, this means by reference to the conception of God that arises out of consideration of Jesus' life, death and resurrection: the idea of God as Trinity. However, if this is to be an authentic interpretation of the evolutionary process that led to *Homo sapiens* and their symbolic, defining minds, it must take the scientific reconstruction of this process of origin with utmost seriousness, though without being confined to thought that can only be contained within a purely scientific perspective.

To clarify the methodological approach followed here, I will draw briefly on the work of the Latin American liberation theologian Clodovis Boff as set out in his *Theology and Praxis* (1987). One of Boff's concerns was to protect liberation theology from the suspicion that it was merely political thought hidden beneath a thin veneer of theological language.[5] He did so by seeking to demonstrate that, although liberation theology commonly tried to understand the political situation using socio-analytic tools (very often of a Marxist hue), such tools did not contribute to theological reflection directly. Rather, however, they contributed to a "text" of a situation that was subsequently "read" using properly theological concepts. In other words, what was sought was not a political theology, but rather a theology *of* the political. Utilizing the work of the Marxist theoretician Louis Althusser, Boff asserted that what defines a discipline is not its subject matter but the system of concepts used to interpret that subject.

In parallel with Boff, what I intend here is a theology *of* evolution and, in particular, a theology *of* evolved *Homo sapiens*. The raw data of the past will be interpreted though the discipline of paleoanthropology to

---

[5] See, for example, the 1984 Vatican "Instruction on Certain Aspects of the 'Theology of Liberation'", accessible from the Vatican's official website: *http://www.vatican.va*.

produce a "text" that can be read in the light of the classic theological concepts of incarnation and Trinity.

This method will enable the construction of theological meaning by attempting to demonstrate a consonance between the being of God and the being of (evolving) creation. This strategy is rooted in the assumption that the origin, purpose and goal of creation are related not merely to God's will and intention (as if these could float free of God's nature) but to the very essence of God's life.

Although this seems counter-intuitive, in the Judeo-Christian tradition ideas about God the creator follow from, rather than precede, notions of God the redeemer. Thus what is believed about the relationship between God and the created world derives from what is believed about God's actions in history to redeem that world. In a paradigmatic example, it appears that for the Hebrew Bible (referred to by Christians as the Old Testament) a conception of God as creator results from asking the question: "How is the God of the Exodus related to the rest of the world?" Accordingly, the primary place where God is made known is in God's acts of salvation. If, therefore, we wish to link the being of God and the nature of creation together as a strategy for discerning the meaning of the latter, we must begin with what is to be discerned of God from his redemptive acts.

If God's actions are to be authentically revealing, then it has to be assumed that God saves the world by being himself, that is, not through actions that are somehow separable from his nature. It would be devastating to the concept of revelation if God were simply a superb actor playing a part in a play so that one could not confidently argue from reported action to inner identity. Although there are cases where people write to the actors who play their favourite soap opera characters as if they were one and the same, a moment's reflection reveals the foolishness of such a belief and tactic. But, if it is desired to argue from God's actions in the world to the way God is in himself, then it must be assumed that, unlike the situation of the actor on television, God's act and God's being are one. More philosophically, God's freedom is expressed in the coincidence of his essence and existence. God does what he is. In adopting this supposition, I am following a long line of argument, traceable from Irenaeus in the 2nd century, through Augustine in the 5th to Karl Barth in the 20th, that seeks to make the doctrine of the Trinity, so revealed, fundamental to theological thought.

This approach to constructing a theological reflection on the evolution of the human brain/mind, means taking issue with one aspect of J. Wentzel van Huyssteen's otherwise remarkable book *Alone in the world? Human*

Uniqueness in Science and Theology (2006). Here, for reasons of prudence as a participant in interdisciplinary dialogue, he forecloses on the use of theological resources that might present themselves as overly intellectualised and abstract. A conspicuous casualty of this approach is the doctrine of the Trinity (*Ibid,* pp.137,142, cf. pp.149,154). Yet this needlessly robs theology of its most distinctive resource, for the concept of the Trinity is, arguably, the highest integrating summary of the Christian intuition concerning the being of God and the basis of creation. Another victim, unsurprisingly, is the doctrine of the incarnation. But this, I will demonstrate, is a powerful tool in holding God and (evolving) creation together in a meaningful way.

The Emergence of the Human Brain/Mind

As Terrance Deacon points out: biologically *Homo sapiens* are just another ape; mentally, by contrast, we are an entirely new kind of creature (1998 p.23). Van Huyssteen agrees: "the story of our species is virtually the story of the growth of the human brain" (2006 p.86). While brain size does not of itself afford an exhaustive account of internal mental function, it does have the immense benefit of being an objectively measurable facet of hominid evolution. One of our most famous predecessors, 3 million year old "Lucy", a member of *Australopithecus afarensis*, had a cranial capacity of roughly 400cc. On average, humans today boast 1350cc (Lewin 1998 p.445). Yet, more significant than sheer size, is the ratio of brain size to body weight, the encephalisation quotient (EQ). Lewin (1998) offers the following average figures: chimpanzee EQ2.0; Australopithecenes EQ2.5; *Homo habilis/rudolfensis* – the first of the *Homo* lineage – (2.5 – 1.8 million years ago) EQ3.1; *Homo ergaster/ erectus* (1.8 million – 300,000 years ago) EQ3.3; and contemporary *Homo sapiens* EQ5.8 (pp. 449f). Indicative of a telling trend as they are, these figures hide three important facts. First, there is wide variation in brain size between individuals of a species: thus, towards the end of their existence, the largest cranial capacity of *Homo erectus* rivalled the smallest of modern humans. Second, by the time of *Homo neanderthalensis* (250,000 – 28,000 years ago) cranial capacity actually slightly exceeded our modern average at 1450cc (Lewin p.366). And third, since the first appearance of *Homo sapiens*, there seems to have been a slight downward trend in brain size. But so much for size! What of the mind within?

Steven Mithen, in his *The Prehistory of the Mind: A Search for the Origin of Art, Science and Religion* (1996; see also Chapter One of this book), makes a lucid and compelling analysis of the possible course of the emergence of the modern human mind. It has to do primarily with the emergence of a particular internal structure. In essence, his thesis concerns a move from separate modal intelligences, that characterised our hominid forebears, to what he terms full "cognitive fluidity", an achievement made exclusively by *Homo sapiens* sometime after their initial emergence roughly 200,000 years ago in sub-Saharan Africa. A consequence and catalyst of this process is language and self-reflexive consciousness. These escape from their originating function in the mode of social intelligence to be the ground for a mind that can examine and integrate thought from its two other modes: knowledge of natural history; and technical practice. The explosive result is a mind that has the capacity to think symbolically.

In Mithen's more recent work, *The Singing Neanderthals: The Origins of Music, Language, Mind and Body* (2005), the argument is taken a stage further. Taking up the neglected topic of the origin of music, Mithen sees in the inter-related though distinct ways in which language and music operate in the brain (2005 Part One) a heuristic clue to earlier forms of hominid communication (Part Two). Along the vector of development that must connect the "grunts, barks and gestures" of monkeys and apes with the fully formed language of contemporary humans, Mithen proposes the emergence of "Hmmmm(m)". This is a holistic (whole phrases, not individual words), multi-modal (voice and gesture), manipulative (rather than referential) and musical form of communication. In *Homo ergaster* and more latterly *Homo heidelbergensis* and *Homo neanderthalensis* it advances to include a mimetic element (for example the sounds and behaviours of a particular prey). Significantly for Mithen, however, it is only with *Homo sapiens*, that, by a process of segmentation, the holistic phrases of "Hmmmmm" are broken into individual words which, through the schematics of grammar, can be combined and recombined in an infinite variety of ways. Only in this do we reach a properly symbolic mentality – the mentality of our own species.

Evidence in the archaeological record that such a symbolic mind has emerged is extensive and persuasive. This is particularly so in Europe where the contrast in manufacture and behaviour between the incoming *Homo sapiens* (known locally as Cro-Magnons) and the existing population of Neanderthals is sharp.[6] The new mentality manifests itself in

---

[6] There is, however, a late-flowering Neanderthal industry (35-30,000 years ago), known as the Chatelperronian. This includes the manufacture of jewellery from bone and teeth using distinctive methods. While Mithen (2005 pp.231f) resolutely refuses to see this as evidence of the independent emergence of symbolic thinking in Neanderthals, preferring to explain it by way of imitation of the incoming *Homo*

the onset of cave art (with which we began), musical instruments, new highly varied tool technologies, sea-voyaging, body decoration and elaborate burial of the dead. Religious sensibility, social stratification and symbolic codification are the legacy of this new way of thinking. However, it is becoming increasingly clear that the inception of this new mentality must have taken place earlier in Africa before *Homo sapiens* spread out from this land of origin to colonize the globe, effectively replacing all other populations of hominid. Mithen (2005, and Chapter One herein) reports the discovery, by Christopher Henshilwood's team, of a 70,000 year old "shale-like ochre" piece inscribed with a cross-hatch design (pp.250f). Found at Blombos cave in South Africa, this is currently the oldest evidence of a symbolic mindset.

Despite the fact that the Neanderthals were the one and only hominid species (of which we are currently aware) fully adapted to the tundra-like European environment[7], 10,000 years after *Homo sapiens* had colonised Europe only this African interloper remained. Zubrow (1989) has calculated that a 2% survival advantage would lead to the extinction of Neanderthals within 1000 years. Those, like Deacon (1998), who are not so convinced of modern human superiority wonder whether disease brought by the incoming population was not more decisive (pp. 370-374).

The Power of Language

Whatever the exact process of its emergence, language, as a symbolic system of words and grammar, shapes minds in such a fundamental way that we cannot over-emphasise its contribution to our identity. As van Huyssteen (2006) rightly asserts: "Language is without doubt the most distinctive human adaptation" (p.234). Language enables engagement in self-conscious thought and reasoning. Ian Tattersall (2000) puts it well:

> [L]anguage is not simply the medium by which we express our ideas and experiences to each other. Rather it is fundamental to the thought process itself....It is in effect impossible for us to conceive of thought (as we are familiar with it) in the absence of language (p.44).

---

sapiens, others disagree. Conway Morris (2003 pp.278-281), for example, sees it as evidence of evolutionary convergence.

[7] Mitochondrial (Mt)DNA analysis from three different and temporally divergent Neanderthal skeletons reveal them as a separate species from *Homo sapiens* and make any significant inter-breeding highly unlikely (Stone and Lurquin, 2007, pp.171-174). MtDNa and Y-chromosome analysis of modern populations also suggest an African origin some time before 150,000 years ago (*Ibid.* pp.174-180).

Language-empowered thought gives us access to an imagined, virtual world. We can ask questions about what might have been and what might be, so granting a sense of past and future. We can contemplate a time before we existed and after we have died (cf. Deacon 1998 p.22). In this way, language contributes to a notion of self, awareness of our own mortality and a conception of the other, of the not-self (cf. van Huyssteen 2006 p.232). Consequently, we can understand ourselves placed in relation to a context of space, time and society, a localised, embodied mind in a world beyond. These abilities give us a survival advantage. As Harry Jerison observes, we are able to construct more useful representations of our material and social worlds (Lewin, 1998 p.464f). These abilities also lie at the root of the religious sensibility that appears to be a universal feature of human understanding and behaviour. Language enables us to conceive of a better world (for example the spirit world of shamanistic religion), to ask questions of ultimate meaning and to posit another "self" in the form of God (Potts 2004 pp.265f). Occupying this exalted epistemological position though is not without its downside. As Arthur Peacocke (1993) observed, our abilities engender a sense of "misfit" between ourselves and our environment. They set us on a consciously unfulfilable quest for self-realisation and the attempt to satisfy aspirations that life cannot meet (p.77). We are both aware of our limits and unwilling to exist within them. For example, we know ourselves to be mortal; yet we yearn for eternity (cf. Eccl 3:11).

Evolution of the Mind: The Enabling Ground of the Incarnation

We turn now to begin the process of theological reflection and start with a consideration of the doctrine of the incarnation. In doing so, it is hoped to open up a way of holding together both the contingency of the evolution of the brain/mind of *Homo sapiens* and its possible meaning.

The "problem" of the incarnation is usually conceived as this: How could a human person, specifically Jesus of Nazareth, also be God? This is the issue that various forms of "liberal" theology have sought to solve by redefining what was intended by Jesus' divinity. The classic approach of Schleiermacher (1928, German original 1831), which has been followed in numerous ways, was to say that Jesus differed from us quantitatively, by degree, not qualitatively. For Schleiermacher it was the unremitting potency of his consciousness of God that set him apart.[8] But the "problem"

---

[8] For a more recent take on this approach see Hick (1993).

of the incarnation could be constructed in a different way: How did there arise a creature that could become the locus of God's life?

Consider one of the key biblical texts that contributes towards a notion of the incarnation.

> In the beginning was the Word [*Logos*], and the Word was with God, and the Word was God ... And the Word became flesh and dwelt among us. (John 1:1,14a)

The "Word" of John's Prologue includes, but goes beyond, the language of God. It reveals a likely indebtedness to Philo of Alexandria (died 45 CE). He was a Hellenistic Jewish thinker who combined the Stoic philosophical notion of the *Logos*, as the pervading rational principle that gives structure and order to the world, with the Hebrew Bible's conception of both "word" and "wisdom" (Dunn 1980 pp.163ff). As the opening chapter of the book of Genesis reveals, the word of God is creative; it therefore does what it says. Significantly, the Hebrew term for word, *debar*, means both word and action. And God's wisdom could already be thought of as a quasi person alongside God, who acts as God's agent in creation and salvation (cf. Proverbs 8:22-31).[9] John inherits, but moves beyond, these ideas. Not only is the Word fully equated with God ("and the Word was God"), yet without simplistically collapsing the Word and God together ("the Word was with God") but also, in a startlingly new move, the claim is made that this Word became flesh, became the human person, Jesus (cf. John 1:17). Thus God acts and communicates through an aspect of his own life that has now become extended to a particular *Homo sapiens*.

Could God have become incarnate before 70,000 years ago, or before the advent of the symbolic power of a human mind most conspicuously evidenced by language?[10] The incarnation has two sides to it. It is both that "the Word became flesh" and also that flesh became the Word. And in order for flesh to become the Word, was it not also necessary for flesh to have become capable of words; that is for there to have emerged a creature capable of language, reason, symbolic thought: and in this way, a creature capable of a relationship with God who is beyond the immediately demonstrable world? Might it therefore be ventured that the enabling ground of the incarnation is also the evolution of the embodied mind of *Homo sapiens*? What is fundamental is the emergence of a particular kind

---

[9] I am grateful here to my colleague Dr Brian Capper.

[10] This is necessarily a speculative question, but still a fruitful thought experiment.

of mind that is not merely more intelligent than its forebears, but of a different intelligence (cf. Tattersall 1998 pp.58ff).

A famous axiom of medieval theology, used by Thomas Aquinas among others, is that: "grace does not destroy, but presupposes and perfects nature". If the grace of God can be said to be operative in the incarnation, then this principle implies that the Word becoming flesh is not an alien imposition on our humanity. Rather, it is the bringing of our humanity to fulfilment (cf. Rahner 1966 p.117). If evolution is a process in which flesh becomes (capable of) words, then it could be argued that the incarnation is the anticipatory[11] fulfilment of evolution, and so a ground of evolution's meaningfulness.[12] From this initial conclusion we can now begin our venture into the realm of contingency.

Whatever else Jesus is, he is an evolved being. His unique humanity also includes his genetic code (the product of a near-4-billion-year process) and a particular developmental history of interaction with his environment. Consequently, the emergence of Jesus' humanity must be conceived as a *contingent* process. There was never a guarantee that precisely this individual, in every circumstance of his existence, would come to be. Jesus' solidarity with us demands no less contingency. Consider the following classic statement concerning the incarnation, offered by the Council of Chalcedon (451 CE).

> ... our Lord Jesus Christ ... truly God and truly man ... one substance with the Father as regards his Godhead, and at the same time of *one substance with us as regards his manhood*; *like us in all respects*, apart from sin ... recognized in two natures, without confusion, *without change*, without division, without separation; the distinction of natures being in no way annulled by the union, but rather *the characteristics of each nature being preserved* and coming together to form one person and subsistence ...[13]

The parts of this "Chalcedonian Definition" that I have italicised are those which ensure Jesus' participation in *our* humanity – our contingently

---

[11] The reason for this qualification will become clear in the concluding section on the redemption of evolution.

[12] Seeking to find the meaning of the evolution of life in the incarnation is not an original move. The pioneer of such attempts was perhaps J.R. Illingworth in his essay for *Lux Mundi* (1904), "The Incarnation in Relation to Development". An important twentieth century contributor to this way of thinking was Teilhard de Chardin. See, for example, his *The Future of Man* (1964).

[13] Bettenson (1944), my emphases.

evolved humanity.[14] According to Gregory of Nazianzus' famous dictum: "What is not assumed is not healed". If Jesus is to be considered our Saviour then he must be like us in all respects except sin.[15] Accordingly, the incarnation becomes a way in which God can be said to embrace, and so value, contingency. This highly significant conclusion means that God is to be found not only in relation to what is necessary and eternal, but also in relation to that which is contingent, open and temporal. For the Christian claim is that God is found in Jesus; and Jesus, as we have seen, is a contingent member of the species *Homo sapiens*.

Thinking about Contingency

Following the challenge of Stephen Jay Gould, thus far it has been assumed that viewing the process of evolution leading to the emergence of *Homo sapiens* as a contingent one equates to seeing our existence as being no more than "the luck of the draw". However, this may not be the case at all.

The contingent (what might not have been) is not the *exact* opposite of the necessary (what had to be). The impression that it is is created, *inter alia*, by the so-called cosmological arguments for God's existence which contrast the dependence of the contingent world on the necessarily (or self-) existing God (as classically found, for example in Aquinas' "Five Ways"[16]). However, the exact opposite of that which is necessary, is that which is random (more biblically the notion of chaos[17]). While randomness is indeed a possible form of contingency, critically it is not its only form.

By way of illustration of the difference, consider the following example. Imagine a snooker table that is perfectly flat, but with no pockets, just a rectangle of continuous cushion. Suppose a player is asked to shoot a ball anywhere he or she desires. Provided the ball does not jump the cushions, every possible resting place on the table has a more or less equal probability. This might model a random process (though, of course, psychologists will take exception to this). Now imagine the same player

---

[14] Many commentators, including Peacocke (1993 pp.275-279), and Macquarrie (1990 pp.392-394), have noted that insisting on Jesus' genetic solidarity with us means that an apparently straight-forward interpretation of the "virgin birth" actually undermines, rather than supports, the incarnation. It makes Jesus seem like a human replica rather than one who shares in the 4-billion-year ancestry of life.
[15] Cf. Hebrews 4:15.
[16] See Aquinas (1964 [1266]).
[17] Cf. Bonting, Chapter Seventeen of this volume, & refs therein. Ed.

and table but, this time, instead of the surface being perfectly flat it is contoured with interconnecting peaks and troughs. This time – although one could not predict the exact location of the ball – its final position would be channelled and constrained, to some extent, by the topography of the table. Some previously possible resting places would be rendered impossible; others will now vary widely in their probability. This models a contingent process that is no longer purely random.

Might something akin to such a channelling of results be applied to the process of evolution? One indication that it might is that the contingent process of the evolution of the brain/mind in *Homo sapiens* is at least susceptible to attempted post *hoc* rational explanation by palaeo-anthropologists. Taking this line of thought further is the work of Simon Conway Morris. This is particularly so of his recent *Life's Solution: Inevitable Humans in a Lonely Universe* (2003) which constitutes a major study of the phenomenon of evolutionary convergence, a phenomenon which he passionately believes has not received the attention it deserves.

Convergence takes place when biologically unrelated (or, more accurately, distantly related) organisms evolve, along separate pathways, towards a similar adaptation. The classic example is often taken to be the evolution of the camera eye in cephalopods (such as the octopus) and vertebrates (such as ourselves) (*Ibid.* pp.151-154). Under Conway Morris' gaze, however, convergence is seen as a ubiquitous occurrence which *Life's Solution* systematically catalogues.[18] Germane to our discussion is his analysis of the degree of convergence that pertains to what might be considered three of the defining marks of *Homo sapiens*: large-brained intelligence; advanced communication; and tool use. He points out that it is less than 2 million years ago, with the advent of *Homo ergaster/erectus*, that the encephalisation quotient of hominids has surpassed that of the dolphins (*Ibid.* p.247). This leads him to suggest that large brains are favoured in environments of particular challenge, such as the dramatic cooling of the Southern Ocean that heralded the ensuing ice age, thus shaping the evolution our own lineage. The persistence of large brains (20 million years for the line that led to dolphins and porpoises), he conjectures, is a response to the emergence of sophisticated social organisation and a corresponding requirement for complex vocalisations (pp.248f). Tool use, and manufacture, is found not only among Chimpanzees, but also New World monkeys (whose lineage has been separate for about 30 million years) and, more surprisingly perhaps, New Caledonian crows (pp.261f). Progress towards something like the same

---

[18] See especially the convergence index (pp.457-461).

mentality has thus occurred in three very different environments: the oceans; the air and the jungle.

This brief summary cannot hope to do justice to the full complexity of Conway Morris' arguments. But, significantly for our purposes, his conclusions are very different from those of Gould that we met earlier. He summarises thus:

> The principal aim of this book has been to show that the constraints of evolution and the ubiquity of convergence make the emergence of something like ourselves a near-inevitability" (p.328).

By this he does not mean that if the process of evolution was re-run from scratch *Homo sapiens* would emerge each time in every exact detail. Rather what is highly likely to emerge is an organism bearing our determinative capacities; one that would not be so different from ourselves. The vision of aliens in the *Star Trek* series, nearly all of which are humanoid, may not be so far wrong (cf. *Ibid.* pp.331f)!

The evolutionary process is characterised by remarkable creative freedom. Human evolution itself points to this irrepressible creativity. So far about 20 species of hominid have been identified, with a peak of diversity about 1.8 million years ago (Tattersall 2000). So-called "adaptive radiation" colours the evolution of us hominids just as much as of the rest of evolved life. There was not a single path of linear development culminating in ourselves, but rather a number of contemporaneously running experiments which included evolutionary dead-ends (such as *Homo neaderthalensis*). But it seems that the phenomenon of contingence provides some grounds at least for thinking that we are far from a complete fluke of history. Rather, it is as if the "bush" of human evolution actually reveals some deeper structure in the nature of life's unfolding.

We might describe this combination of openness and constraint in the evolution of *Homo sapiens* as "freedom within a form". This will assist our subsequent theological reflections. In itself, however, this broad conclusion is not new. Arthur Peacocke (1993), for example, spoke about the interplay of chance and law that is necessary in a creatively constructive process (pp.65f; cf. p.245). And, following the language of Popper, he was prepared to speak of "propensities" in the evolutionary process, such as that for increased complexity. Such propensities, he argued, would, given enough time, inevitably lead to the emergence of self-consciousness (pp.220f).

Before bringing these reflections on contingency to a close, it is worth drawing attention to how vital contingency is to our human worth. It makes our existence both precarious (I might not have existed) and meaningful (what I choose to do makes a difference). Without an open future, our lives would be either fully determined, or else merely a trial run, in time, at attaining an eternal and immutable identity already assigned to us by God. Neither view would be an incentive to get up and start the new day. This freedom, which contingency confers, and that renders our decisions meaningful, we must note, is also a "freedom within a form". Each of us possesses a unique, unfolding and developing field of freedom. It is defined by our genetic inheritance, early development, life opportunities and previous choices. It is the field of freedom that helps to delimit our unfurling identity.

Freedom within a Form: A Vestige of the Trinity

Earlier, I suggested that one route to the construction of theological meaning lay via the discernment of a consonance between the being of God and the being of the world. Further, following Barth, I assumed that God's being and God's action are one. Combining these positions implies a strategy in which it is asked whether there exists a divine correspondent to the "freedom within a form" that we found characterised the evolutionary process which led to *Homo sapiens*. More precisely, can the deep structure of life's evolutionary unfolding, revealed in the phenomenon of convergence, be grounded in the nature of God? I now wish to propose that just such a connection can be found with the Trinitarian pattern of God's life.

The Trinitarian life of God could be seen as one of the free play of creative love in infinite freedom, yet expressed within the specific form of the relationships between the three Persons of Father, Son and Holy Spirit. In part, this is a conception expressed in the doctrine of "perichoresis". Stemming from the Greek *perichorein* (which means to penetrate, to contain, or to envelope) the doctrine of perichoresis points to the mutual interpenetration of Trinitarian Persons in complete freedom yet without the abolition of the distinct identity of each Person (Turner 1983). It is a conception that speaks both of love and unity. Its biblical roots lie in the Gospel of John where Jesus talks about his being in the Father and the Father in him. Moreover, this is a form of unity which is open for the world's eventual inclusion (John 17:20-23).

During the mediaeval period it was common to understand this relationship as a kind of divine dance, perhaps because of the similarity of the term *perichoreuein*.[19] This can be seen as a way of expressing the

---

[19] This observation is made by Robinson (2004).

notion that, far from being static, the divine life has its own movement and dynamic. The image of a dance, of course, corresponds exactly with what we are looking for, something that has both definite form and yet an infinite variety of (contingent) possible expression.

This notion of perichoresis, however, can be of more use. It can help link together what we have begun to say about the Trinitarian life of God and our earlier application of the doctrine of the incarnation. It can do so because of its history. The idea of perichoresis first emerges in Eastern theology as a way of understanding the incarnation. In the 7th century Maximus the Confessor used it to speak of the mutual indwelling of the human and divine natures in Christ. Shortly afterwards, John of Damascus applied it to the Trinitarian life of God (Turner 1983). If perichoresis can be re-described as an over-reaching of the one into the other then the connection becomes clear. The incarnation can be conceived as an analogy of the pattern of the divine life itself. There is a "naturalness" about the way in which a God whose inner life consists in the over-reaching of one Person into another could express that life through an over-reaching of the divine life into the world. Thus the incarnation itself can be thought of as an action which flows from God's Trinitarian being. We therefore note that the incarnation can be considered as grounded in the "freedom within a form" of God's life.

Trinity as Conversation

What I am attempting to do is to set up a pattern of inter-locking analogies that can relate four things together: the Trinitarian being of God; the incarnation; the form of human distinctiveness (as capable of words); and the evolutionary process by which that distinctiveness arises. Such a set of analogies, I wish to claim, can grant theological meaning to humanity's contingent evolution. To do this more effectively I now seek to unfold the Trinitarian life of God as one of conversation.[20] In this way we can add greater precision to the broader concept of "freedom within a form".

In the garden of Bishop's Palace in Exeter is a sculpture by Frances Favata entitled "Trinity". It consists of three separate figures, clearly relating together, with space between them for the observer to walk and so join the apparent conversation. How might this suggestive insight into the Trinity be grounded theologically?

---

[20] This notion is not new. Two other recent developments of its promise are those of Jenson (1997) and Schwöbel (2003).

I want to use someone who may seem an unlikely ally, Karl Barth. Barth is most commonly associated with what might be termed "totalitarian revelation". This follows from his desire to unfold the Trinitarian being of God from God's absolute lordship in revelation. God is in control of every aspect of revelation; he is the initiating subject, the content and the very possibility of this content being successfully received. Accordingly Barth declares,

> Thus it is God Himself, it is the same God in His unimpeded unity, who according to the biblical understanding of revelation is the revealing God, the event of revelation and its effect on man (1975 p.309)

Yet, hidden away in *Church Dogmatics I.1* is another view of revelation. It follows from Barth's reflections on what it means for the Holy Spirit to enable us to be accounted children of God. Here Barth states:

> In receiving the Holy Ghost [a person] is what in himself and of himself he cannot be, one who belongs to God as a child to its father, one who knows God as a child knows its father (*Ibid.* p.457)

Barth seems to be implying that the Spirit enables us to know God by setting us, as adopted children, within earshot, as it were, of God's own internal conversation of love. The Spirit, as the medium of the eternal conversation between Father and Son, of their communion together, acts in time to include us in that same conversation (cf. *Ibid,* p.471). As far as I am aware, Barth does not actually use the word "conversation" to convey his meaning; but his intimations in this direction are what I want to take up and briefly develop.

The notion of God's internal life as a conversation of love could find some support in the New Testament. Father and Son engage in conversation (John 12:27-29). Even if the Son is primarily associated with the Word (*Logos*), as we saw above, certainly the Father (Matthew 3:16-17; Mark 9:7) and the Spirit (Matthew 10:18-20) are also connected to words. Additionally, the Spirit is portrayed as searching and thus revealing the thoughts (and so words) of God's inner life (1 Corinthians 2: 10b-13; cf. John 16:13-15).

Conversation would also appear to be an apt metaphor for the Trinitarian life. Conversation, at its best, is rooted in an assumption of equality and a pattern of reciprocity[21]. It is less about an exchange of

---

[21] There is not space within the scope of this paper to engage with the vital question of human evolution and sin. However, it would be possible to advance the

information and more about the building up of a shared, common life. It subsists in freedom and is motivated by joy in an exploration of the other via shared experience. In this way, conversation points to a relational notion of identity. One only comes to a true sense of self through, and not apart from, the other.

The main motivation, however, for developing this notion of Trinity as conversation is that it will assist in the task of laying out the set of interlocking analogies, adduced above, that will help afford theological meaning to the evolution of the human brain/mind.

First, conversation is an excellent image for God's perichoretic life. In conversation there is a sharing of the contents of minds of the participants. The thoughts of one come to indwell the mind of the other. Conversation also affirms a sense of dynamic relationship, infinitely fertile in its own creativity and rooted in love[22]. As such it resonates with Moltmann's (1980) "social doctrine of the Trinity".

Secondly, if conversation is seen as the ability to overreach oneself into the other, then, as earlier argued, it becomes an analogue of the incarnation: of the unfurling of the eternal divine life into the world of space and time. The Word of the divine life becomes flesh as the Trinitarian conversation of God opens itself to the world. In this way, then, we also preserve the connection between God's being and God's act.

Thirdly, there is a profound consonance between a Trinitarian God of conversation and a world that evolves into words in the distinctive emergent abilities of *Homo sapiens*. Evolution thus opens up the possibility that the words of the world and the Word of God can become part of the same conversation. The primary place for the realisation of this possibility is the incarnate *Homo sapiens*, Jesus. This proposal is strengthened by a facet of the deep structure of evolution revealed in convergence. There is a strong correlation found, not only in the primates, but also in the cetaceans, between size of brain and size of social group (Lewin 1998 pp.452-454). When formidable intelligence emerges from the

---

notion of "bad conversation" (a term suggested to me by Timothy Gorringe) as an illuminating route into this issue. Words (and actions) used to defend self-interest and self-advantage, to manipulate and deceive, to cause unnecessary pain and suffering, to render others into instruments of one's own interest rather than promote a community of equality – all these would be manifestations of such sinful bad conversation (cf. James 3). [See also Chapter Fifteen. Ed.]

[22] Trinity as conversation could, therefore, be a route to affirming both: that God's existence is radically independent from creation – God does not need creation to have something to talk about; and also essentially open for creation – there is room for creation to subsist within God's essential conversation.

evolutionary process, it appears to be first and foremost social intelligence. It emerges as the ability to understand, predict and manipulate the actions of others. Intelligence arises, therefore, as essentially social intelligence: intelligence capable of conversation.

Fourthly, the contingent process of evolution, its unpredictability in detail yet its sense of destination, as seen in the phenomenon of convergence – this "freedom within a form" – can find its correlative in the being of God. The Trinitarian conversation is also a "freedom within a form". Good conversation is always contingent conversation. It cannot be the uttering of the same thing over and over again. Yet it also has a reciprocal, turn-taking shape that prevents it from degrading into a monologue. This approach could be deepened by the work of Robinson (2004) who, drawing on the philosophy of Charles Peirce, has discerned a profound consonance between the interpretive, semiotic, processes of all life and the Trinitarian being of God.

Last, the notion of conversation protects the distinguishable identity of the participants, preventing them from being merged together in an undifferentiated whole. We now have a tool that will enable us to conceive of the world participating in God's life, anticipated in the incarnation, without the world thereby losing its identity as non-God. There is, thus, no "Borg-type" assimilation. Such an ability to differentiate between God and world is important because of the parallels between the proposals I have made here and the philosophy of Hegel[23].

Hegel conceives of ultimate reality as Absolute Spirit/Mind (*Geist*). This is not a static reality but one that advances by means of a dialectical process of thesis, antithesis and synthesis. Thus, Absolute Spirit (thesis) objectifies itself through projection as nature (antithesis) This, then, advances (in a dialectical manner) to the point of the emergence of human consciousness (finite spirit). Through finite spirit's recognition of itself as one with Absolute Spirit, Absolute Spirit finally comes to itself at a new level of self-understanding (synthesis). Compare this with the suggestions made here. Evolution is a process by which flesh becomes (capable of) words as an enabling condition for the Word to become flesh. This then opens the way for an inclusive exchange between the words of creation and the Trinitarian God of conversation. Hegel conceives of a process in which matter becomes mind which is open for Mind, i.e. of a process in which matter becomes words which are open for the Word.

One of the difficulties of Hegel's system is that its explanatory power (he claims to uncover the basic rationality of all reality) is won at the price

---

[23] An admirably clear introduction to Hegel's philosophy is Copleston's (1963).

of turning history into a necessary process (a problem inherited by Marxism). A second difficulty is that he collapses the process of the world into the process of God (pantheism, or more generously panentheism[24]). It is hoped that a set of interlocking analogies rooted in the metaphor of conversation can escape these traps, both by allowing sufficient room for contingency on the one hand, and by underlining the distinguishable identities of God and the (evolving) world on the other. Not to do so is to run the risk of seeing evolution as a necessary process which gains theological meaning only as it is reduced to an aspect of God's own life, so losing an individual and intrinsic meaning of its own. Rather, we have sought to assert evolution's meaning precisely through comprehension of its contingent unfolding. It would be wrong, therefore, to read the emergence of *Homo sapiens* from sub-Saharan Africa, and their subsequent replacement of all other hominid populations, as if this were theologically fore-ordained. This would make *Homo sapiens* God's chosen species over all others: "Out of Africa I called my child" (cf. Hosea 11:1). It is preferable to see evolution as contingently moving in the direction of words and to understand ourselves, with more humility, as the locus of the emergence of this property.

The Image of God and the Community of Creation

If the story of evolution can be legitimately conceived as a process in which flesh, and through flesh ultimately matter, becomes (capable of) words, then, theologically viewed, this can be seen as a process orientated on the incarnation. While wishing to preserve the unique sense in which *the* incarnation is applied to Jesus alone, might not this road of development be signposted by other "incarnations" along the way? This extended language of incarnation probably makes more sense if it is related to the motif of "the image of God" – the *Imago Dei* (Genesis 1:26).

While van Huyssteen (2006) might be accused of being rather thin in his treatment of the theological motifs of incarnation and Trinity, this is certainly not the case when it comes to the *Imago Dei*. It functions as his main theological resource for comprehending the unique abilities of *Homo sapiens* as revealed by the various scientific disciplines that bear on human evolution. Thus he argues that, "the theological notion of the *Imago Dei* [can be] powerfully revisioned as *emerging from nature itself*. For the theologian, this interdisciplinary move implies that God used natural

---

[24] This is the notion that the world is included in, but does not exhaustively constitute, the being of God.

history for religion and for religious belief to emerge as a natural phenomenon" (p.322). However, this leads him to restrict the *Imago Dei*'s application to *Homo sapiens* alone on the basis of our uniqueness (*Ibid.*).

To say that God's image can be found in the world is, as a minimum, surely to say that there exists a worldly locus of God's representation, and so a form of God's presence. But must this notion be restricted to *Homo sapiens* as has been traditionally the case?

Mithen's (2005) explorations into the evolutionary predecessor of language, Hmmmm(m), were briefly outlined above. In part they stem from an attempt to explain evidence for social cooperation (notable in hunting) and cohesion in the archaeological record that points to an effective communication system of the past. One emotionally compelling strand of that evidence is the indication of care for the sick. Conway Morris (2003) reports possible skeletal evidence for care of a member of *Homo erectus* suffering from an overdose of vitamin A, most probably sourced from carnivore liver (p.274). A particularly striking example comes from a Neanderthal skeleton found in the Shanidar Cave in Iraq where, presumably through the attention of other members of the social group, this individual was able to survive blindness, head injuries and a crushing of the right-hand side of the body that possibly resulted from a cave collapse (Mithen 1996 p.134). Increasing social cohesion may have resulted in part as the by-product of more effective bipedalism, a process that begins with *Homo ergaster* close to 2 million years ago (Mithen 2005 pp.144f). Narrower hips, which improve walking efficiency, mean that the brain can only reach a certain limited stage of development before an infant is born. Thus infants remain highly dependent on their mothers while further rapid brain development takes place. This, in turn, necessitates their mothers' reliance upon the social cohesion of the group (cf. *Ibid.* pp.184f).

Might not this evidence of growing social cohesion, and the presumed emergence of more sophisticated versions of Hmmmm(m), be interpreted as a development of the image of God even before the distinctive cognitive fluidity of later *Homo sapiens* renders religious consciousness and awareness explicitly possible?[25] Two particular considerations make this more plausible. First, the motif of the "image of God" has always had to contend with a differentiation; that between ourselves and the perfect image, Jesus Christ (cf. 2 Corinthians 4:4; Colossians 1:15). Gradation is

---

[25] Barth's (1966 §45.2) connection between a relational conception of the *Imago Dei* and the fundamentally social nature of humanity, "being as encounter", could be deployed to strengthen this point.

thus conceivable. Second, by linking human evolution and the Trinity together through the metaphor of conversation, it is possible to see the development of the brain/mind not so much in terms of the acquisition of abstract prowess (for example the acquisition of symbolic culture) but as indicating the growth of relational behaviour and thinking. Indeed, "cognitive fluidity" might itself be seen as a highly developed form of relationality. It is worth remembering at this juncture that the incarnation of the Word made flesh is not just about words, it is also, supremely about "body language", the expression of divine life in the concrete shape of a human life. This adds particular significance to the gestural elements of the presumed Hmmmm(m) form of communication. A false sense of our distinctiveness is suggested by the contemporary gap in ability between chimpanzees (our closest living genetic relative) and ourselves. We, however, need to remember the intervening rungs of the evolutionary ladder that led up to us which significantly narrow this comforting distinction. In other words, not to extend the language of *Imago Dei* in some form to the whole *Homo* lineage is a false constraint of a creationist perspective.

This notion of a growth in the image of God through the evolutionary process emphasises both the connectedness of the community of creation, important for our ethical response to the natural world, and the somatic connection between God's saving action in Jesus and the rest of evolved creation, important for thinking about the redemption of evolution. It is to this latter consideration that we now turn by way of conclusion.

The Redemption of Evolution

I have tried to unfold the possible consonance between a creation that evolves into words and the being of God. In this way evolution can function as a category of the creative intention of God. What evolution cannot be, however, is a category of life's redemption. This is because, as Moltmann (1990) observes, evolution takes place in time (p.302). The death of individuals, that drives the evolution of life, is just one facet of the ontological problem of the transience of creation. Our experience of time is one of irreversibility. The open possibilities of the future are selectively realised in the moving moment of the present before being set hard as unalterable past. An event once happened – a deed once done – cannot be undone. The imperfections, incompleteness and suffering of the past thus become stubborn, petrified questions that no future prospect, however marvellous, can answer. The Christian claim is that the only place from which an answer can be found is the eschatological future which is both revealed and grounded in the resurrection of Jesus from the dead. The eschatological future is thus nothing less than a transformation of the nature of time (cf. Moltmann 1995 p.26). It means that every once-present moment can be liberated from its captivity in unalterable past, brought into company with every other and together transformed through the possibilities of God's infinite life.

Twice earlier I have spoken of the incarnation as the anticipation of the fulfilment of (evolving) creation. In the incarnation human words and the divine Word are united in a new level of conversation. But when this Jesus is raised from death into the (dynamic) eternity of God's life, his evolutionary solidarity with us, by way of his humanity, becomes the gateway in principle for the raising of all evolved life into God's living conversation[26]. It is through the efficacy of the Holy Spirit that each individual element of creation can access Jesus as this gateway to inclusion in the divine conversation. Thus, to take a personal favourite hominid predecessor, *Homo neanderthalensis*, Neanderthals are decisively not a mere working sketch to be consigned to the dustbin of history as inherently surpassable. They rather are lost strands of conversation waiting to be taken up again by God. Justly proud as we are of our evolved brain/mind, salvation cannot, like a university education, be restricted only to those who have demonstrated the intellectual ability to benefit from it. God's conversation with the world is far more encapsulating than mere philosophy.[27]

References

Aquinas, T. 1964. *Summa Theologiae* Vol. 2 (Ia. 2-11), ed. T. McDermott. London: Blackmans.
Bailey, R. 2001. "Overcoming Veriphobia – Learning to love truth again", *British Journal of Educational Studies*, Vol 49/2 159-172.
Barth, K. 1975. *Church Dogmatics* I.1. Edinburgh: T & T Clark.
—. 1960. *Church Dogmatics* III.2. Edinburgh: T & T Clark.
Bettenson, H. 1944. *Documents of the Christian Church*, London: Humphrey Milford.

---

[26] Conversation thus also becomes a metaphor for salvation. Cf. Jesus' words: "No longer do I call you servants, for the servant does not know what his master is doing; but I have called you friends, for all that I have heard from my Father I have made known to you" (John 15:15).

[27] I am grateful to my colleague Dr Terrance Clifford-Amos whose numerous suggestions have helped improve the clarity of this piece.

Boff, C. 1987. *Theology and Praxis*, Maryknoll: Orbis.
Copleston, F. 1963. *A History of Philosophy* Vol. VII. London: Search Press.
Conway Morris, S. 1998. *The Crucible of Creation: The Burgess Shale and the Rise of Animals*. Oxford: University Press.
—. 2003. *Life's Solution: Inevitable Humans in a Lonely Universe*. Cambridge: University Press.
de Chardin, P. Teilhard. 1969. *The Future of Man*. London: Fontana.
Deacon, T. 1998. *The Symbolic Species: The Co-evolution of Language and the Human Brain*. London: Penguin.
Dennett, D. 1993. *Consciousness Explained*. London: Penguin.
Dunn, J.D.G.. 1980. *Christology in the Making*. London: SCM.
Forrest, B. 2000. "The Possibility of Meaning in Human Evolution". Zygon 35 (4), 861-880.
Gadamer, H.-G. 1975. *Truth and Method*. London: Sheed & Ward.
Gould, S.J. 1980. *Ever Since Darwin*. London: Penguin.
—. 1996. *Life's Grandeur: The Spread of Excellence from Plato to Darwin*. London: Random House.
Hick, J. 1993. *The Myth of God Incarnate*, 2nd edn. London: SCM.
Illingworth, J.R.. 1904. "The Incarnation in relation to development", in *Lux Mundi: a series of studies in the religion of the Incarnation*, ed.C. Gore. London: John Murray.
Jenson, R.W. 1997. *Systematic Theology* Vol. 1*: The Triune God*. Oxford: University Press.
Lewin, R. 1998. *Principles of Human Evolution: A Core Textbook*. London: Blackwell Science.
Macquarrie, J. 1990. *Jesus Christ in Modern Thought*. London: SCM.
McKie, R. 2000. *Ape Man*. London: BBC.
Mithen, S.J. 1996. *The Prehistory of the Mind: A Search for the Origin of Art, Science and Religion*. London: Thames & Hudson.
—. 2005. *The Singing Neanderthals: The Origins of Music, Language, Mind and Body*. London: Phoenix.
Moltmann, J. 1985. *God in Creation: An Ecological Doctrine of Creation*. London: SCM.
—. 1990. *The Way of Jesus Christ: Christology in Messianic Dimensions*. London: SCM.
Peacocke, A.R.. 1993. *Theology for a Scientific Age*. London: SCM.
Potts, R. 2004. Sociality and the concept of culture in human origins, in *The Origins and Nature of Sociality (*R.W Sussman & A.R. Chapman, eds). New York: Walter de Gruyter.

Rahner, K. 1966. *Theological Investigations* Vol. 4. London: Darton, Longman & Todd.
Robinson, A.J. 2004. C.S. Peirce as a resource for a theology of evolution. *Zygon* 39 (1), 111-136.
Schleiermacher, F. 1928. *The Christian Faith* (edn 2, H. R. Mackintosh & J. S. Stewart eds). Edinburgh: T & T Clark.
Schwöbel, C. 2003. God as conversation: Reflections on a theological ontology of communicative relations, in *Theology and Conversation: Towards a Relational Theology* (J. Hares & P. De May, eds). Leuven: Peeters Publishers.
Southgate, C. 2008. *The Groaning of Creation: God, Evolution and the Problem of Evil*. London: Westminster John Knox.
Stone, L. & Lurquin, P.F. with Cavalli-Sforza, L.L.. 2007. *Genes, Culture, and Human Evolution: A Synthesis*. Oxford: Blackwell.
Tattersall, I. 1998. *Becoming Human: Evolution and Human Uniqueness*. New York: Harcourt Brace.
—. 2000. Once we were not alone. *Scientific American* vol 282 (1) (Jan), 56-62.
Turner, H. E. W. 1983. Coinherence., in *A New Dictionary of Christian Theology*, (A. Richardson & J. Bowden eds). London: SCM.
van Huyssteen, W.J. 2006. *Alone in the World? Human Uniqueness in Science and Theology*. Grand Rapids: Eerdmans.
Zubrow, E. 1989. The demographic modelling of Neanderthal extinction, in *The Human Revolution*, (P. Mellars & C. Stringer eds). Edinburgh: University Press.

CHAPTER ELEVEN

RESPONSE TO JEREMY LAW

ROGER KNIGHT

Revd Roger Knight studied science at school, was for two years a medical student at the Middlesex Hospital Medical School before opting instead for theology. He subsequently graduated BD, AKC from King's College, London and was ordained as an Anglican priest. He has been Rector of Cuxton and Halling in the Diocese of Rochester since 1987. The 2007 Conference also marked the end of a period in which he had served as the SRF's Secretary.

His response to Jeremy Law focuses on the latter's stress on evolutionary contingency. Extending this theme, Roger Knight explores the paradoxical relationship, in Biblical thought and in theology, between contingency and Divine will. His talk is reprinted here almost exactly as delivered at the 2007 meeting.

The quest for meaning

One day last week I was summoned to the telephone with the words, "It's someone from the Science and Fiction Forum." If we were to equate religion with fiction, would meaning still be a problem? As Dr Law says, the quest for meaning plays a determinative part in the history of human intellectual endeavour. Religious people, who believe that the universe is the creation of a rational mind, analogous to theirs, though infinitely greater, logically expect to find meaning in the way things are, and are therefore susceptible to being troubled by evidence of contingency, chance, happenstance and randomness. An atheist might attempt to explain our longing for meaning in terms of our evolutionary survival. For some reason, beings who seek meaning in life generally live longer and produce more offspring than those who do not. Longing for meaning may be a heritable characteristic with implications for survival and reproduction.

Perhaps there is no need for meaning actually to mean anything. The evolutionary point is that the *quest* for meaning has survival benefits. The theist might respond, however, that the reason why questing for meaning has survival benefits may be that existence does in fact ultimately have meaning!

An agnostic might say that we cannot possibly know whether our lives and the universe itself have meaning and that therefore it is useless to ask the question. The agnostic, however, cannot expect to be left in peace by the ardent theist who believes that the concept of God is not merely an answer to an intellectual puzzle, but that God is someone to whom we can and should relate, someone who commands us to live in a certain way and someone on whom we depend for what happens to us in time and eternity. Neither can the agnostic expect peace from the passionate atheist, who believes that religion is a force for evil, hindering humanity's advance by its insistence on the authority of outmoded and flawed beliefs about the universe, how it works and what its purpose is, and the underlying cause of far too many wars and persecutions. The passion of such an atheist, however, surely derives from his strong sense of what it means to be human?

So the question of meaning does not go away if we are not religious. Yet any religion, based on belief in an omnipotent and omniscient Deity, is in difficulties when confronted with modern Science, which finds contingency in everything. Not only are chance mutation and survival the driving force of evolution, but Chaos Theory forces us to accept the essential unpredictability of our environment, while quantum uncertainties are apparently fundamental characteristics of the very stuff of which everything is made. The ubiquity of contingency was, of course, as upsetting to classical Science as it is to classical Christian Theology, prompting Einstein's oft-quoted insistence, "God does not play dice!"

Dr Law offers us a way of welcoming contingency as part of our quest for meaning within the context of faith in God. He speaks very much from within the Christian tradition, but perhaps the insights he provides may be humbly offered to those of other faiths, as well as to those of no faith, as pointers to the way in which we may ultimately resolve the paradox of meaning and contingency.

Necessity and contingency

It is interesting that Dr Law turns to biblical accounts of Redemption, rather than Creation, pointing out that the biblical account of God as Creator grows out of the experience of his redemptive love for his people.

These redemption stories have always implicitly, and sometimes explicitly, raised the question of necessity and contingency. The Egyptian Pharaoh and the kings of Assyria, Babylon and Persia are treated in the Hebrew Scriptures as morally responsible individuals. They make decisions for which they are answerable. Yet they are also portrayed as effectively the tools of God. The decisions they take inevitably glorify God, punish transgressions and ultimately set God's people free. The way the Bible tells the story, it is not merely that omniscience necessarily knows in advance what Pharaoh and other kings will do; their actions are actually determined by a God, who is very much more powerful than they are. God tells Moses to demand of Pharaoh that he set the Israelite slaves free, but he also says,

> I will harden Pharaoh's heart, and multiply my signs and my wonders in the land of Egypt. But Pharaoh shall not hearken unto you, that I may lay my hand upon Egypt, and bring forth mine armies and my people the children of Israel, out of the land of Egypt by great judgments. (Exodus 7: 3-4)

We seem to be expected to believe simultaneously that Pharaoh's actions are contingent (they could have been otherwise) and that what happened was inevitable. It is surely the experience of the Exodus which gives meaning to the existence of the Jewish people and this is not, in the biblical tradition, a meaning only discovered, or perhaps created, after the event, with the benefit of hindsight, but already determined in the counsels of God, when he called Abraham's family to leave Ur of the Chaldees, centuries before the Israelites migrated to Egypt. Maybe Moses and the prophets stand accused of a naïve faith, characterised by its ability to believe 100 impossible things before breakfast, but perhaps instead they have a sophistication and a subtlety regarding contingency and meaning which eludes us. How could men who lived two to three thousand years ago have subtler insights than we have? Classically, Jews, Christians and Moslems would reply that they received special inspiration from God. This explanation is not available to the academic historian or scientist, who follows Hume in discounting the miraculous as an explanation of anything, but it may be open to us if we accept Dr Law's account of the possibility of human beings being somehow included in the inner life of the Godhead.

(It is worth observing here that it is not only in the Old Testament that we are confronted with this clash of contingency and necessity. Judas Iscariot apparently necessarily plays his pre-ordained part in the story of the Crucifixion, but he is also regarded as responsible for his own actions. Jesus says:

> The Son of man goeth as it is written of him: but woe unto that man by whom the Son of man is betrayed! It had been good for that man if he had not been born. (Matthew 26: 24)

More generally, the New Testament invites men and women to choose to follow Jesus, but insists that God has chosen those who do decide to take up his invitation and that they could not have chosen God unless he had given them the grace to do so!)

Dr Law relates contingency and meaning in the universe with their meeting within the very nature of God. He can do this in the Christian tradition supremely because of the doctrine of the Incarnation, though this too has two very serious problems with regard to necessity and contingency. Could Mary have refused the angel's invitation to become the Mother of God's Son? If she had, would there have been no Incarnation or is it absurd to think that God might have had other virgins in reserve, just in case – or maybe even that the Mary we know was not the first choice? More seriously, Christianity classically believes that the human nature of Jesus was in all respects the same as ours, except without sin. Christians have always insisted that his human nature received no assistance from his divine nature in resisting temptation. It has always been one of the difficulties that we are required to believe that the second person of the Trinity became one person with the only human being who has ever lived who has not shared the sinful nature of the archetypal Adam. It seems to me to raise even greater difficulties if we are to believe that the human nature of Jesus was the result of an almost infinite number of contingent events, between the Big Bang and the reign of King Herod the Great.

Maybe, as Dr Law says, following Simon Conway Morris, the evolution of something like humanity was inevitable, given the very nature of the universe – a nature which must derive from the nature of the Creator God, if Law is correct – but the evolution of any given individual could never be guaranteed. This seems to be a very strong version of the Anthropic Principle, that the universe was so created that humanity would evolve. I suppose that believers in an infinitely powerful God could develop a super-strong version of the Anthropic Principle by which the universe would have been so set up that all the individuals who exist and who have ever existed, inevitably exist; but this would remove all contingency from the universe and make the appearance of both chance and choice mere illusions.

In the Hebrew Scriptures it is the Word of God which effects the creation and it is through God's Word, spoken by his servants the prophets, that God controls human history. The Christian faith in the Incarnation is indeed that Jesus is the Word made flesh. It is elegant to link *emperichoresis* – human beings being caught up in the relationships within the Trinity – with the idea of conversation, but it is hard to see how the universe develops to such a level of sophistication unless it is designed so to develop, or unless there is divine intervention (occasional or constant) to guide its development.

As I read Dr Law's paper, I found myself wondering whether radioactive decay might be a helpful analogy? As I understand it, it is impossible to know when a given nucleus will decay. In that sense every time, say, a Radium nucleus becomes a Thorium nucleus, it is a contingent event. It might not have happened. It is possible to know the half life, i.e. to know when half of the nuclei will have decayed, yet not which half! This seems to me to make it possible for the nuclear physicist to be like the Old Testament prophet and believe that phenomena are simultaneously contingent and necessary – or maybe, if Hegel is useful to us, that necessity and contingency are thesis and antithesis, looking for a synthesis in a great mind, perhaps the Mind of God.

Necessary Resurrection – a paradox?

I cannot conclude without saying that I am puzzled by Dr Law's introduction of the Resurrection as necessary to the final consummation for which "the whole creation groaneth and travaileth in pain until now" (Romans 8: 22). Secular Evolutionary Theory, with all the problems it raises for Theology because of its insistence on chance and contingency, is founded on the Enlightenment belief, stated so well by David Hume, that it is unprofitable to speculate about the possibility of miraculous interventions. We are always to assume that Miracle stories are based on deception or on the misinterpretation of phenomena which actually have a natural explanation. It is this rationalist approach which has made modern science so successful and given it its huge authority relative to other disciplines, despite the fact that science's philosophical underpinning is no more secure than that of religion or of any other human quest for meaning. If we allow resurrection as part of the story, why not allow other divine interventions in the story of evolution? Could not God have lent a helping hand to the first lung fish to emerge from the waters? Could not God have given an injection of consciousness to hominids when they reached a stage of development that nature alone could never improve on? If we allow divine intervention whenever it is required to keep evolution on track to produce beings capable of relating to God, the problem and the opportunity of contingency both disappear. Personally, I can believe in a God Who runs the Universe and Who not only foresees the coming of a Messiah of the seed of David but of Whom it can be said that Jesus and everything which happens to him is part of his "determinate counsel and foreknowledge" (Acts 2: 23). I can also understand those who either do not believe in God or else believe that God does not intervene in the way the universe unfolds, still less in the lives of individuals. But, with apologies to Dr Law, I cannot see why we have to explain almost everything on the assumption of divine non-intervention, and then look to God for a decisive intervention in order to make sense of it all in the end.

CHAPTER TWELVE

THEOLOGY, EVOLUTION AND THE MIND

ROGER PAUL

Revd Dr Roger Paul read Social Anthropology at Cambridge, and almost 25 years later (1998) took an Open University PhD with a thesis on the interaction between an insect herbivore and its plant host (1998). In between, he had read Divinity, also at Cambridge, and been ordained into the Anglican priesthood in 1981. His last parish post was as Rector of Kirkby Stephen (Carlisle Diocese); he is now National Adviser (Unity in Mission), Council for Christian Unity, Archbishops' Council. He also provides invaluable service to the SRF as its current Treasurer.

In the final paper of the conference, expanded below, Dr Paul summarised and drew together the five Main Papers which had gone before. But he did more – he moved beyond them, to a near-Hegelian synthesis which makes his paper, to the Editor's mind, the most important in this volume.

Introduction

In this final chapter, I shall explore some of the themes that weave their way through the preceding chapters. Inevitably, with contributions from a number of disciplines and philosophical viewpoints, there is a healthy diversity of ideas and some lively controversy to pick over. Here, I discuss whether broad evolutionary theory can support theism, and examine Watt's claim that evolutionary theory can liberate theological insight. I then go on to consider the conflicting views of naturalistic evolutionary epistemology, as expounded by Spurway, and dualism, commended by Trigg, and the relationship betweens brains and minds. I then reflect on the nature of religious metaphor, central to Mithen's thesis on the origins of religion, with respect to what we can know or say of God, and discuss the ways in which the science and religion dialogue may be conducted, with special reference to Law. These are just a few of the issues touched upon by the major contributors to this book, and in a short chapter, it has proved a stimulating, though perhaps ultimately impossible, task to bring them together into a semblance of a whole.

Watts 1: A direction for evolution?

One area of controversy that the preceding chapters touch upon is whether evolution is open to the continuing guidance of a creator God. Fraser Watts (Chapter Four) argues that there is evidence for directional change in evolution towards more complex organisms, which are able to process information about their environment more efficiently. He avoids identifying directional change with any notion of progress, but does suggest that the evolution of organisms, such as human beings, with what we may call a spiritual consciousness, is highly significant. He questions the "narrow evolutionary theorists" (his phrase), who emphasise the process of evolution as one of pure chance, which would discount such openness. In contrast, Watts discusses the work of Simon Conway Morris (1988, 2006), whose concept of evolutionary convergence points to common features of organisms that have independently evolved, suggesting that indeed some other process or influence than natural selection is at work. What this process or influence may be, Conway-Morris leaves intentionally vague, but he does allow that evolutionary convergence may be an indication of purpose, or divine guidance.

It would be possible to argue then that evolutionary directional change suggests a *telos* or purpose in evolution: that a Final Cause is moving evolution in the direction of emerging life forms that are capable of consciousness, rational thought and moral judgements. This is close to speaking of the divine creative activity, working through the processes of evolution and resulting in the emergence of such life forms (Ward 1996). Although proponents of this essentially theistic view seek to avoid speaking of special divine action, but more of a general divine guidance, there is a short step from this view to some of the claims of intelligent design, where divine intention, direct intervention and special action feature prominently.

However, the ideas of Stuart Kauffman (1989, 2002), mentioned in passing by Watts, deserve closer attention in this discussion. Certainly, Kauffman attempts to identify laws of complexity that explain how complex, autonomous organisms can emerge. One feature of complex systems is that the range of possible states, into which an actual system can move – what Kauffman calls "the adjacent possible" – increases as the complexity of the system increases. He suggests that a system expands the scale of its adjacent possible, on average, as rapidly as it can. The counterbalance to this expansion is the need for systems to persist in time, and therefore to display a level of stability. A system that is too open will be too unstable, while a system that is too closed will have little opportunity for development. Kauffman suggests that emergent systems are therefore to be found on the "edge of chaos".

Kauffman assumes that the transition from what is possible to what actually emerges is an irreversible process, but how do we understand that transition? Analytical, scientific models can be run forwards or backwards in time, so for example, Newton's Laws of Motion can be applied to determine the state of a body at some time in the future, and equally can be used to establish its state at some previous time, if all the conditions of the body can at some time be pre-stated. Another feature of complex systems, according to Kauffman, is that the persistent novelty, which is the "doings, the embodied know-how, the carryings on of autonomous agents" within the system, cannot be pre-stated. The laws of complexity that Kauffman is searching for are therefore not time symmetrical. Mathematically, such laws cannot be expressed analytically, and are limited to *post hoc* descriptions, or at best simulations.

As Watts points out, these ideas are by no means tested scientifically, and are at present interesting speculations based on a patchy body of evidence. However, the key point for our discussion is that Kauffman is putting forward essentially scientific arguments. Even if laws of complexity are identifiable, they do not necessarily support theism; they can still be understood wholly in terms of natural law. Even if evolutionary directional change to ever increasing levels of complexity proves to be supported scientifically, we must be careful about extrapolating from the scientific hypothesis to theological statements. The temptation is to focus on the scientific understanding, not always the most orthodox, and to hold it up as supporting a particular theological viewpoint. To be fair to Watts, he is careful to refer to God as, in a general sense, working through evolution in terms of opening up our understanding of creation as ongoing rather than as a once and for all divine act, and he sees this as the key insight that evolutionary theory has brought to theology.

Watts 2: Salvation and incarnation

Watts, then, encourages Christians, and perhaps people of other faiths as well, to embrace the liberation that broad evolutionary theory brings to our theological understanding. His aim is to ask how evolutionary theory can illuminate certain aspects of Christian doctrine, which may be refined and developed in the light of that theory, such as the doctrine of creation, sin and the fall, the incarnation and the nature of salvation. To work towards reconciling evolutionary thought with these issues is an important task for theology, and as Watts shows, helps to serve as a corrective where doctrine has come adrift from the best understanding we have of human nature and the emergence of species, and the biological processes which are at work. For Christians his discussion of salvation history is particularly pertinent, setting it within the overall development of moral consciousness in humans, which is associated with biological evolution. The Biblical narratives can then be interpreted within this framework. So the narrative of the fall in Genesis 3 offers an insight into the capacity of humans to do deliberate evil, which is the darker side of the raising of moral consciousness. In a similar way, the incarnation of Christ is another instance of "a turning point in the development of human consciousness". Within this reading Christ makes possible a new consciousness of the indwelling spirit within humanity.

A number of fundamental theological issues remain in the light of this revision of salvation history. What does it mean for example that Christ is both human and divine? Watts sees Christ as emerging within the processes of evolution, quoting Gerd Theissen (1984) in his suggestion that Christ marks a new "spiritual mutation". One can reconcile Christ's emergence with the creative work of God, which as Watts suggests, is to be identified generally with the on-going process of evolution. This interpretation, however, does not answer the question of the distinctiveness of Christ, because all humans, indeed all creatures may be open to the creative work of God. If the normative way for God to work in creation is through the processes of evolution, then to maintain the distinctiveness of one event requires a special, distinctive input. Why we should regard God as working in Christ more significantly than in any other instance of the evolution of consciousness needs to be more fully explored, and we may well be led back into the clutches of supernaturalism.[1]

Mithen and cognitive fluidity

Steven Mithen's hypothesis (Chapter One and 1996, 2006) of the development of separate modular intelligences in early *homo* species,

---

[1] Note the similarity of this comment to Roger Knight's last sentence (Chapter Eleven). Ed.

which become linked as language develops, giving rise to cognitive fluidity, suggests a developmental framework, in which the emergence of consciousness can be understood. He discusses this process in relation to detailed archaeological evidence, especially by comparing and contrasting early *Homo sapiens* with other contemporary *homo* species, especially *Homo neanderthalis*. Mithen suggests that the cognitive fluidity of early humans, emerging around about 100,000 years ago, that makes possible metaphorical language and thought, thereby providing the basis for art, religion and science, gave them a survival advantage. In contrast, Neanderthals, without this cognitive fluidity, and therefore not developing advanced, creative technological, scientific and artistic skills, struggled to survive in the harsh environments of the ice age. The hypothesis neatly explains the "big bang" of cultural expansion of early humans in the archaeological record between 70 and 30,000 years ago. There are issues here about the testing of this hypothesis – a hypothesis that is framed at least in part from the archaeological record, and tested by the same data. There is also the danger of projecting back our modern preconceptions about how early humans thought onto an obscure distant past. There is the possibility, as Fraser Watts suggests, that language was metaphorical from the outset – that language is inherently metaphorical. From his view, metaphorical thought, and therefore religion and art, would not have emerged as a result of the emergence of cognitive fluidity, but with the development of language itself.

However, there are also issues about the very process of evolution that is envisaged in the evolution of the mind. Darwinian Evolution by natural selection is by definition genetic and biological so, strictly speaking, we can only speak of the Darwinian evolution of the brain by natural selection. A selection advantage is gained by, say, increased brain size, as happened first with early hominids about 2 million years ago, and later in early humans about 500,000 years ago. Increased brain size made possible the living in larger groups, which may have given an advantage in harsher environments. Likewise, the development of the neuro-physiological basis of language would give a further selective advantage. It is important to note that the evolution of the brain in these ways could not happen without certain other physiological characteristics evolving, such as efficient bi-pedalism, and changes in breathing physiology and the larynx. There is also evidence that other specific modules of intelligence as well as linguistic ability may be located in specific areas of the brain.

Spurway's argument from evolutionary epistemology

The mind evolves in a Darwinian sense only insofar as it is dependent on the structure of the brain. There is of course an intimate relation between mind and brain, and so a biological explanation for the evolution of the mind may go quite a long way. In his exposition of evolutionary epistemology (Chapter Six), Neil Spurway gives an essentially biological, physical explanation of the development of human cognition, in which perceptions and conceptions are both tested in the competition for survival. An organism's perception of the world through the sense organs is highly honed to the environment in which it has to make a living, and Spurway gives a number of telling examples of this process of adaptation. So for example, the range of wavelength of light that the human eye can perceive is such because perceiving within that range has been proved successful for survival. Our spatial awareness also will have conferred a selection advantage at some stage in our evolutionary history. When it comes to mental capacities, concepts and cognitive thought in general, evolutionary epistemology argues that these will also have been tested out in the struggle for survival. Spurway thus refers to biological propensities evolving within a species' *umwelt*, or the context, life situation in which a species has to get along, thrive, adapt and make a living. The *umwelt* is the arena of biological natural selection, which tests out the properties of a species. Thus in human life we can trust our biological nature, because it has been tested, it has in this way a truth content within our *umwelt*. In this respect our cognitive capacities, as well as our sensory capabilities, have been tried, tested, and found to work in our world. If they did not, we would not be here.

It is important to recognise, however, that mental capacities are not determined in detail by neurological and physical evolution. Spurway is careful to write of propensities, rather than the detail of mental capacities, evolving biologically. We may then think of a linguistic propensity, which enables an infant to learn a particular language, given the context and relationships within which that learning can take place. A child is not born with the ability to speak German or any other particular language, but has the toolbox to learn whatever language the child is exposed to.

Although Spurway does not reject the worlds of subjective feelings, artistic expression and the religious quest as meaningless, there is a sense in which these areas of human life are considered to be dependent on the more fundamental reality of the biological. Here we have to be aware of the possibility of other possible influences in the way that the mind has developed, particularly social and cultural influences. Some would wish to question the extent to which evolutionary psychology may speak of cultural evolution. If evolution is by selection, what are the mechanisms of this selection? Can we speak of cultural adaptation to new situations, environments, and knowledge, as evolution? Cultural change is a complex interaction of individual creativity, ecological and environmental factors, and communication between different human groups. The processes of cultural transmission, enculturation, education and nurture are vital elements in driving cultural change. How far they can be accommodated within an evolutionary paradigm remains open to question.

Trigg's case for dualism

Roger Trigg (Chapter Eight and 2002) poses an indirect challenge to Spurway's position, presenting a dualistic view in which both physical and mental, objective and subjective, matter and spirit are equally important and valid. He accepts the reality of the physical and biological, but argues for the recognition of the equally autonomous reality of that which is not physical: the subjective reality of the mind, soul and spirit. So, for example, we can move between a subjective view (say the experience of pain) to an objective view, which looks at the physical causes of pain (Trigg 1970). A doctor treating the pain will focus primarily on the medical diagnosis, but will have an eye also on the subjective experience of the patient. One of the first questions one is asked at Accident and Emergency is "On a scale of 1 to 10, how painful is your injury?" In a similar way, we can move between a consideration of the brain and a consideration of mind, two very different perspectives. But can we say that one perspective is more fundamental than the other? Trigg's answer is to say that each is fundamental and irreducible, and without the perspective provided by each, we will be missing something in our understanding of the world. He goes on to argue that there is a correspondence between the dualism of body and self, and of the dualism between the world and God, and that in a fundamental way, if we give ground on one dualism, we also lose the other. Theism, then, rests on a dualistic understanding of reality, and once accepted gives a very powerful account of why the world is as it is and of the nature of human beings. Roger Trigg wants there to be space for both scientific and theological discourse to take place, without one posing a threat to the other, and he does indeed carve out a space equally for body and soul, world and God, to be taken seriously in his apologetic for dualism.

There are many issues in Trigg's position. It is encouraging to the theologian that theological discourse has its own valid space, but is this space gained at the expense of being isolated from other discourses, especially the scientific? If there is a domain in which it is proper to engage in theology, does this imply that in the scientific domain theology has no place? This question could easily be asked the other way round as well. Behind it is the more philosophical issue of the relationship between the two worlds: how do body and soul interact? How does God engage with the world? Trigg acknowledges these questions as "the perennial issue" of dualism, but unfortunately he does not explore that issue further.

Emergence, and multiple perspectives

Is it possible to treat these issues in a non-dualist, and naturalistic way? The notion of emergence, as a property of structures and organisms of ever-increasing complexity, with many different embedded levels of organisation, may be one approach. Of crucial importance is the nature of the relation between emergent properties and their constituent lower levels of organisation. If the mind is an emergent property of the brain, how do the two relate to one another? One model of this relation is that of supervenience (Drai 1999). Strong supervenience supposes that an event in the brain, say the firing of a network of neurons, has a direct and particular expression in terms of the activity of the mind, whereas weak supervenience supposes a general dependence of mind on brain, such that the mind is embodied in the brain, and does not persist in an independent way. Beyond supervenience, another model of emergence could be that an emergent property has the potential to "cut loose" from its substrate and have some sort of autonomous existence, giving rise to the possibility of top-down causality, in which an emergent property acts on lower levels of organisation (Murphy 2002). Trigg accepts that emergence may provide a scientific paradigm for the development of the brain, but argues that it does not give an adequate philosophical account of the self-conscious mind, because the subjective thoughts and feelings of the mind are irreducible.

An alternative view, neither dualist nor relying on the notion of emergence, is to think of multiple perspectives on one reality, a reality, which is complex and defying an understanding in terms of one fundamental explanation (Ward 1996). The reality with which we have to do is very complex and subtle, and we, beings with very complex brains and minds constantly need to seek explanations and meanings in different ways. It is not only possible, but also important for us to move between these different perspectives and allow them to illuminate each other. Mary Midgely (Midgley 1998), for example, applies the notion of different mental maps, which can be used to understand and explain different aspects of the same terrain. Problems occur when we take one map as the only definitive perspective, as with reductionism, because it excludes elements of the reality we are trying to understand.

The importance of metaphorical thought

Metaphorical thought and language have an important place in the emergence and development of religion. Both Mithen and Spurway broadly agree that the origin of religion is bound up with the emergence of the capacity or propensity for metaphorical thought, which, for Spurway, is neurologically hard wired. Without this propensity, religion could not be expressed but the precise expression is not determined by the biological propensity. Unfortunately, this theory is applied only in a general sense, in terms of the cognitive capabilities of humans in general. It may be important, however, to look more closely at the differences between individuals within the same culture, and also between cultures. The psychology of types, for example, pioneered by C. G. Jung (Jung 1971) and developed by Isobel Briggs Myers (Myers 1995) indicates that the human mind has a range of different functions. Individuals may be stronger in certain types of function and weaker in others, for example in logical thought compared to differentiation of feelings, and will have a unique, dynamic combination of types, which describes personality. The application of this theory of types to individuals' religious responses indicates a variety of approaches to religious faith and practice, and may help to explain why some people appear to be more open to religious symbolism than others. It is possible that the differences between individuals are the result of differences in neurological hardwiring, as well as in individuals' development. The propensity for metaphorical thought in the human mind may be associated more with certain personality types than others, and this calls into question the universality of metaphorical thought as the basis of religion.

Mithen focuses particularly on the imagist expression of religion, rather than the doctrinal. The former emphasises more what religion does, its function, rather than its truth content as a statement about the nature of reality. The religious imagination uses very potent images drawn from experience but applied to the supernatural, or spiritual, domain, in order to express a whole range of emotions and needs of ultimate meaning, value and consequence. These images, essentially metaphorical, are the medium through which religion gains expression. Mithen agrees that religion must have some selection advantage, but does not develop that theme. It would be possible to speak in terms of bonding of human groupings, or as offering a sense of place and belonging. The function of religious symbols, likewise, may have a unifying function, and be important in the raising of consciousness of the connectedness of things. The functions of religion are many and divers, and will include sociological, psychological and moral functions.

In contrast, doctrinal religion is concerned more directly with truth content. Spurway argues that unlike those capacities, which have been tested by natural selection, concepts of religion, which are metaphysical concepts, do not relate to our *umwelt*. They are not testable therefore through natural selection, and it is questionable whether they can be trusted. He suggests that these concepts extend intuitive knowledge and poetic thought, into metaphysical entities, the truth content of which we cannot prove. I am not so sure, however, that the distinction between metaphor and metaphysics is quite so clear. The question is whether the choice of metaphor is arbitrary, or whether there is an inner connection between the metaphor and what it refers to, and here it is important to make some finer distinctions. The first line of one of the sonnets of Gerard Manley Hopkins contains two metaphors that invite the reader to reflect imaginatively:

"As kingfishers catch fire, dragonflies draw flame" (Hopkins 1948)

Hopkins is not, obviously, writing literally here. First, we hear the sounds of the words, particularly the alliteration between the metaphors and what they describe. In these metaphors Hopkins is helping us to reflect on the qualities of the flight of these two creatures. A kingfisher darts along the course of a river, and often what one notices first is the blue flash of wings, as if it were catching fire. A dragonfly hovers with rapid wings almost imperceptible, as if it were fanning flames into life. Neither kingfishers nor dragonflies literally burn up as they fly, and Hopkins has no intention of encouraging us to think that, but something in their flight suggests a quality of fire or flame. A metaphor works in poetry and literature if there is a connection with what it is referring to, yet without identification. There must be tension between the dissimilarity, which distinguishes the metaphor from what it describes, and the similarity, which connects them together. We may speak here of the tension between consonance and dissonance.

When we reflect on religious metaphor, we have to tread even more carefully. To say: "God is love", is to say something very important about ultimate values and meaning. But our use of the word "love" is based on its use with respect to human relationships, both the experience of love in its many varied expressions, but equally important our experience of what is not love. To say "God is love" is to speak metaphorically; in other words, in a non-literal sense, yet with a degree of consonance as well as dissonance between the use of the word love referring to human experience, and its reference to God (Davies 2002). The theological justification for the application of this and other metaphors to God has been expressed in various ways. Thomas Aquinas argued that theological language is not a private language (Kerr 2002), but uses words, which share their meanings with other contexts of speech. The first point is that the word love, for example must carry something of the same meaning whether it refers to human experience or to God. However, we would not say that a particular person "is love", because human love is always qualified and limited, and is a distinct quality or attribute that may or may not be observed in a person's actions. With respect to God, Thomas Aquinas argues in Summa Theologia (trans 1911) that all that God is, is intrinsic to God's very being. To use metaphors in this way is to use them analogically, and not in an arbitrary or figurative way. This does not guarantee, however, the truth content of theological statements, and at this point Spurway is right to challenge religious dogmatism. However, Aquinas understands that because God is the reason for, or First Cause of, there being something rather than nothing, all that exists participates in the being of God, and has the potential to express something of God in an analogical way.

Law 1: The incarnation of the Word

Jeremy Law (Chapter Ten) expresses this idea in terms of seeking a "consonance between the being of (evolving) creation and the being of God". His essay can then be understood as an exercise in developing an appropriate evolutionary metaphor in which to speak of God. His methodology is to make connections between themes within the Christian theological tradition concerning the Holy Trinity and insights drawn from evolutionary theory. His starting point then is to reflect on the incarnation as God embracing and valuing contingency, by virtue of the full, complete humanity of Jesus, who is like us in every respect. Like us, "Jesus is an evolved being". He argues that in order for the Word to become flesh, the flesh (organism) had to become capable of words, in other words to be capable of rational thought and language. Later in his essay, Law relates this to the idea of God's image being capable of representation in the world. For the incarnation, then, to be possible, in Law's view, an organism capable of representing God's nature had to evolve. Evolution may then be viewed as the process in which such an organism, not necessarily human, may emerge. But as we have already seen, according to Thomas Aquinas, all that exists is sustained by God because it participates in God's being. If this is the case, all that exists has the potentiality to express (at least something of) God's nature, irrespective of whether it is capable of words or not. Law himself, in referring to Philo's notion of the Word or *logos,* seems to suggest that it is the rational principle that orders the world. All that exists must participate in this rational principle, simply by being in existence. In order for Law's thesis to stand, we need to introduce a notion of degrees of representation of God, such that in the incarnation, Christ is the fullest representation that is available to us. There is Biblical support for this view in Hebrews, where the ways in which God spoke through the prophets is contrasted with the incarnation, in which:

> [the Son] is the reflection of God's glory and the exact imprint of God's very being (Hebrews 1:3).

Although there is a sense of finality here, the incarnation is not a unique revelation of God. Towards the end of his essay, Law does express this view by speaking of the image of God being represented through many stages in evolutionary history, and suggesting that as evolution unfolds, the image grows. While this idea relies heavily on a sense of progress which, as we discussed earlier, and as Fraser Watts pointed out, is not wholly supportable, it does suggest the solidarity of humans, traditionally considered as the exclusive locus of the image of God, with the whole stream of life.

Law 2: Contingency and direction

Crucially, for Law, evolution is open and contingent. He argues, however, for evolution also to be constrained, in a way similar to that which Watts discussed in his essay, drawing particularly on the work on evolutionary convergence of Simon Conway Morris. Contingency does not necessarily mean randomness, but does suggest dependency on what actually happens and openness in terms of outcome, while at the same time being susceptible of being channelled in certain ways with discernable *post hoc* patterns emerging. Law's central concept with respect to (contingent) evolution is "freedom within a form", and he offers the images of the movement of a billiard ball within the cushions of the table, and of dance, music and conversation to express this idea.

The consonance that Law tries to identify with evolutionary "freedom within a form" is with the theological concept of *emperichoresis*, which refers to the mutual interpenetration, or indwelling, of the persons of the Trinity, which yet retains each person's identity and distinction. For Law, this idea suggests the creative freedom and interplay of love within the life of the Trinity. Contingency, as "freedom within a form", is thus present, within the dynamic of the creative interplay between the persons of the Trinity. He develops the analogy of conversation between the persons of the Trinity, by considering the open-ended, exploratory nature of conversation, yet with an unfolding pattern that progresses to a resolution. However, I do not think that this means that we must admit that contingency is intrinsic to God's being. The qualities of creative freedom and loving interplay in the relationships of the Trinity make sense as qualities of God's necessary being. Unlike an organism, which is at is because of contingent events that might not have happened acting on it, God is not an object or material that can be acted upon and formed by external events. God's actions or energies are inseparable from his essence, so God necessarily does what he is. What God does is not dependent on anything other than God's nature.

The difficulty with Law's thesis, as I understand it, is that he relies heavily on the notion of evolution as proceeding in a certain direction, with the result that ideas of progress and purpose are implicit in his account of evolution. These ideas introduce concepts that are alien to evolutionary theory and perhaps extrapolate too far even from Conway-Morris's notion of evolutionary convergence. Here, I want to point out the caution that Aquinas offered in respect of doing theology:

> They hold a plainly false opinion who say that in regard to the truth of religion it does not matter what a person thinks about the creation as long as he or she has the correct opinion concerning God. An error concerning the creation ends as false thinking about God. (*Summa Contra Gentiles*, trans. 1923)

It is quite legitimate for Law to explore the implications for theology of a controversial scientific idea, provided he acknowledges that an alternative scientific view might radically change the understanding of God that develops. It would be valuable to explore the implications for theology of a narrow interpretation of evolution: would consonance be so easy to come by then? By engaging in different pathways of exploration, the conditional and partial nature of theological understanding becomes apparent. Indeed, we may use two of the metaphors that Law draws out in relation to evolution, to describe such an approach to theology. We may say first that it is contingent on our interpretation of scientific knowledge and theory. In other words, it matters how we interpret the science or, to put more generally, how we understand the way the world is, because this will influence the metaphors we use to speak of God. Secondly, theology is a multi-faceted conversation: between the foundation religious texts, the theological tradition, the sources of knowledge and understanding of the world, and the exercise of reason, logic and faith. Moreover, it is an endeavour that takes place within a social and historical context giving scope for creative, open-ended dialogue. Theology, then, has the potential for playful interaction, while its methods and insights are constantly being tested within the communities of faith, and against the sources of scripture, tradition and contemporary knowledge.

Mysticism and mind's self-transcendence

If our theological understanding and therefore all our conceptions of God, are contingent, what may be said about theological truth? Our position on this will depend on whether we hold a strong or weak view of scientific reductionism. The contributors to this volume display a wide variation on this matter, from the evolutionary epistemological view of Spurway, to Trigg's defence and commendation of nuanced dualism. It is a pity that Spurway does not develop his conclusion, that we need to approach religious belief through the *via negativa*, with humility and wordless awe, and without claiming absolute truth for religious doctrine. There is a dynamic between the affirmative way of understanding God – through symbol, metaphor, words – and the negative way which places these understandings in the greater context of the divine darkness and radical incomprehensibility and wholly-otherness of God (Paul 2006). In this respect, speaking metaphorically of God is radically different to speaking metaphorically about things, which we can perceive, and concepts, which we can conceive. If God is incomprehensible, insensible, inconceivable, then the metaphors we use of God are applied to an unknown. It seems apt to quote Philip Toynbee in this respect:

> We constantly use metaphor in our everyday speech, but we use it legitimately only in cases where we can, if challenged, translate the metaphor back into a more literal and prosaic language ..... The difficulty – some would say the absurdity – of trying to write about the metaphysical realm is that because we can talk about it in no language other than the metaphorical, we can never deliver a literal equivalent but only offer, if challenged, an alternative metaphor." (Toynbee 1973 p.63)

As the writer of the Mystical Theology concluded we have to keep saying: "God is not that", in the same breath that we say "God is that" (Dionysius the Areopagite 1920). On the surface, this appears very similar to the postmodernist way of thinking, in which truth and the absolute become absorbed in the void, and meaning and significance are relativised in the context in which they are expressed (Williams 2007 chaps 1, 2). However, at the heart of the theological tradition of the *via negativa* is the mystical experience of being taken out of oneself in an encounter in the divine darkness, the cloud of unknowing (Lossky 1957). Within this mystical theological tradition, God, although unknowable, is not to be identified with the void.

Evolutionary epistemology, supported by the archaeological record and developmental paradigm that Steven Mithen brings to bear, has great explanatory power, with profound implications for our self-understanding. It roots our species in evolutionary history and demonstrates that the way we think, talk and perceive has emerged out of the struggles for survival of our ancestors. However, there is one question that still nags. The human mind has the immense capacity to reach beyond the world of experience and sense, to transcend itself and so to think thoughts, dream dreams, imagine worlds, and crucially to turn back self-consciously on itself and the world so as to understand itself. It is as if the evolution of mind has led to the capacity for transcendence, freeing the mind from its own evolutionary history. So the question is: How can an evolved mind transcend itself? How can it go beyond the structures, ways of thinking and language that the forces of natural selection have shaped? I hope that this book, and the contributions from such a variety of disciplines and convictions, will stimulate further exploration of this question.

References

Aquinas, T. 1911 edn. *Summa Theologia*, I.a, trans. Fathers of the English Dominican Province. London: Burns, Oates & Washbourne.
—. 1923 edn. *Summa Contra Gentiles*, II.3, trans. Fathers of the English Dominican Province. London: Burns, Oates and Washbourne.
Davies, B. 2002. *Aquinas*. Oxford: Continuum
Dionysius the Areopagite 1920 edn. *The Divine Names and the Mystical Theology*, Rolt, C.E. (trans). London: S.P.C.K.
Drai, D. 1999. *Supervenience and Realism*. Aldershot: Ashgate.
Hopkins, G.M. 1948. *Poems, 3rd Edition*. Oxford: Oxford University Press

Jung, C.G. 1971. *Psychological Types:* Collected Works, Vol. 6. Baynes, H.G. (trans) revised by Hull, R.F.C.. London: Routledge & Kegan Paul.
Kauffman, S. 1995. *At Home in the Universe.* London, Viking.
—. 2000. *Investigations.* Oxford: University Press.
Kerr, F. 2002. *After Aquinas,* Oxford: Blackwell.
Lossky, V. 1957. *The Mystical Theology of the Eastern Church.* London: James Clarke.
Midgley, M.1998. One World, but a Big One, in Rose, R. (ed.) *Brains to Consciousness? Essays on the new sciences of the mind.* London: Penquin,
Mithen, S.J. 1996. *The Prehistory of the Mind: A Search for the Origin of Art, Science and Religion.* London: Thames and Hudson.
—. 2005. *The Singing Neanderthals: The Origins of Music, Language, Mind and Body.* London: Weidenfield & Nicholson.
Morris, S.C. 1998. *The Crucible of Creation.* Oxford: University Press.
—. 2006. *The Boyle Lecture:* Darwin's Compass: How Evolution Discovers the Song of Creation. *Science and Christian Belief* 18 (1) 5-22.
Myers, I. 1995. *Gifts differing: Understanding personality type.* U.S: Davies-Black Publishing.
Murphy, N. 2002. The Problem of Mental Causation: How does reason get its grip on the brain? *Science and Christian Belief,* 14(2) 143-158.
Paul, R.P. 2006. Subjectivist-Observing and Objective-Participant Perspectives on the World: Kant, Aquinas and Quantum Mechanics. *Theology and Science,* 4(2) 151-170.
Theissen, G. 1984. *Biblical Faith: an Evolutionary Approach.* London: SCM.
Toynbee, P. 1973. *Towards the Holy Spirit.* London: SCM.
Trigg, R. 1970. *Pain and Emotion.* Oxford: University Press.
—. 2002. *Philosophy Matters.* Oxford: Blackwell.
Ward, K. 1996. *God, Chance and Necessity.* Oxford: Oneworld.
Williams, R. 2007. *Wrestling with Angels,* London: SCM.

PART 2

CHAPTER THIRTEEN

COSMIC CONVERSATION:
THE EVOLVING DIALOGUE IN MATHEMATICS
BETWEEN MIND AND REALITY

GAVIN HITCHCOCK

Dr Hitchcock read mathematics at Oxford and took a PhD in the field of topology at Keele. He is now a Senior Lecturer in Mathematics, University of Zimbabwe, where he teaches topology, analysis and the history of mathematics. His most recent publication is a book co-authored with a past student, aimed at nurturing potential mathematicians: "A Primer for Mathematics Competitions" (2008). Earlier productions include a film script on calculus and a play about the struggles of 17th and 18th century mathematicians to come to terms with new number concepts.

Readers will share the Editor's sense of privilege and inspiration as, early in this paper, Dr Hitchcock tells of spreading his discipline far from the lecture room, in what is presently one of the most tortured societies in Africa. Later, more orthodoxly yet no less impressively, he spells out the challenge which the extraordinary fit of mathematics to the world presents, for anyone aspiring to explain the evolution of human intellectual capacity.

1. Humans as problem posing, problem solving animals

A central feature of being human is our love of riddles and puzzles, our capacity and penchant for curiosity, questioning, wondering – in short, the human imperative to pose and to solve problems. The best illustration of this, for me, is a memorable afternoon spent interacting with a group of about thirty informal (and illegal) miners seated on a large anthill in the

Zimbabwe bush, engaged in a "workshop on creative problem solving". I first developed such workshops for mixed groups of teachers, and adapted them for specialist mathematics teachers, for groups of children, for special-needs associations, for groups of workers from mining or commercial sectors, and finally for company middle management seminars. It constantly amazed me how problems and inventive solution strategies arising from mathematical contexts could engage attention and elicit a similar quality of enjoyment in such disparate groups. Now, in a new experiment, I found myself in a context far removed from teachers and managers.

The mathematics was there in what we were doing, but cloaked in the form of kinship riddles, matchstick problems, pebble problems, scissors-and-string problems, beads on wire, sharing food, betting on horses, etc. The puzzle-loving element of our common humanity is epitomised by the eager responses of those miners to challenge: practising teamwork naturally, drawing connections quickly, making analogies, seeing for themselves the beauty of simple and unexpected solutions, faces lighting up at the experience of a "Eureka!" moment, one or two even dancing with delight. This workshop group was no different in essence from my other groups, educated or uneducated, younger or older.

For these impoverished men and their dysfunctional communities in a benighted country with the lowest life expectancy and the highest inflation rate in the world, the capacity for creative problem solving has obvious survival value. However, unusually creative or gifted individuals in Africa and elsewhere have traditionally been feared and envied, condemned as witches or sorcerers, and often killed. Totalitarian societies tend to reject novelty or genius as subversive. It is the inquisitive children (and animals) that get into most trouble, the creative child that causes most disruption in classrooms where silence and regimented order are valued. Even in modern times many education systems have neglected, if not repressed, the most gifted children. The human propensity to wonder at the world, and to frame and grapple with far-reaching questions about the world, seems to have transcended any genetic determination and must be investigated as part of cultural evolution. But it is not a simple matter to understand the evolution of such capacities.

2. Science and mathematics in human culture

Within recorded history, cultural context has been very important for certain distinctive expressions of this propensity in mathematical terms. The ancient Babylonians (and also the Egyptians, Indians, Chinese and

Mayans, in strikingly different ways) achieved remarkable sophistication in astronomy, mensuration, manipulating numbers and translating practical problems into geometric diagrams or algebraic equations, often posing and solving problems far transcending any immediate practical utility. Drawing from the achievements of Babylonians and Egyptians, the Greek philosopher-mathematicians of the period 600-300 B.C.E. are widely credited with elevating mathematics to a new dignity as the bridge between the physical, changing temporal world and the unchanging, eternal world of ideas. To an unprecedented degree, they posed and struggled to solve problems that bore no relation whatsoever to utility, beaurocratic rule, military prowess, survival or wealth. And they stated and proved theorems of a purely intellectual character, connected by logical chains of reasoning that have come to characterise what we now call pure mathematics. Medieval Islamic mathematicians, while embracing and further developing aspects of Greek and Indian mathematics, made major contributions of quite different kinds, for example, in algebra, in spherical trigonometry, and in the scientific study and artistic application of two-dimensional symmetry.

A fundamental feature of the gradual transition in Europe (11th to 17th centuries) that helped to spawn what is called the "scientific revolution", was a transformation in attitude to what we now call the natural world, then newly celebrated and respected as "nature" – a transformation from passive awe to active curiosity, and from timid, cowed acceptance to bold, eager enquiry. It has been widely accepted, at least since Whitehead (1926), that this transformation had its roots in the medieval theology of the early universities, incorporating the so-called cultural mandate in the first few chapters of the Book of Genesis, and the versatile, universal metaphors of keeping the garden, and naming the animals. The Adam (humankind) are mandated, delegated, empowered and challenged by an omnipotent yet loving Elohim-Jahweh, who affirms his handiwork as very good, to explore, study, master, care for and delight in this handiwork.

To appreciate what a difference world-view, "control beliefs", or theological presuppositions can make, Africa provides an instructive case study. From a post-Christian, post-scientific perspective, much that goes on in Africa is enigmatic and deeply puzzling. With rich natural resources and human potential, the continent yet struggles to extricate itself from a morass of poverty, disease, ethnic strife and brutal despotism. The causes are complex and diverse, but this predicament derives from something deeper than past colonial injustices and present global trade malpractices. Where animism and the traditional reverence for ancestral spirits is pervasive, nature is feared – indeed it is not perceived as a unified "natural" domain at all in the Western scientific sense, and the rigidly ritualised hierarchical obligations and dependencies of the tribe and extended family can crush any creative, entrepreneurial drives. Human senses are distrusted, individual decision-making and personal initiative is discouraged, and knowledge is not readily shared. The "natural" principles of uniformity and causality, and the "humanist" principles of individual rationality, dignity and responsibility, and the consequent belief in the comprehensibility of the physical world, are not part of the world-view. Imported schemes, methods and machines tend to break down, as a "maintenance" culture is alien, and a truly scientific frame of mind in relation to nature and the technological exploitation of nature is slow to take root. But other valuable aspects of human life in its more social or spiritual dimensions are perhaps more fully expressed in Africa than in the West, and African visual arts, music, literature and dance are vibrant and unsurpassed. And that individual, innate, universal human creative intelligence is there. What we are confronted with in Africa is the big question of what constitutes a *fruitful scientific community* engaging in scientific converse with nature, and what cultural contexts and philosophical presuppositions – what qualities and kinds of faith – will nurture and sustain such communities.

3. The heart of mathematics is problems-plus-solutions

Three academic disciplines exemplify the human problem-posing and problem-solving imperative in their dealing with worlds remote from and apparently (at first) irrelevant to everyday "practical" concerns. Cosmology studies the very big, nuclear physics and quantum theory study the very small, and mathematics studies the very abstract. Mathematics is the prime exemplar, for a mathematician lives to pose and to solve problems. Results are expressed as theorems and proofs, framed in an abstract mental world. At the dawn of the scientific revolution Galileo and Kepler were pioneers in applying mathematics to the study of nature, but the mathematics they applied was over two thousand years old, created in attempts to solve purely mathematical problems. The history of mathematics shows that the lure and fertility of great problems is the main driving force of mathematics.

Many historians of mathematics will agree that the most important legacy bequeathed to us by the ancient Greeks was *open problems*. Here are five of them:

1. To find all "perfect numbers" – numbers like **6=1+2+3** and **28=1+2+4+7+14**, which are equal to the sum of their proper divisors.
2. To construct (with straight edge and compass only) the side of a cube with double the volume of a given cube – that is, to construct the cube root of 2.
3. To trisect a given angle.
4. To construct a regular 7-gon.
5. To construct a square equal in area to a given circle.

In spite of its venerable age (about 2,400 years) the first of these is still not fully solved. (Are there any odd ones? Are there an infinite number of them?) And the other four were resolved relatively recently (as late as the 1872 in the case of the last), by proving each to be impossible with Euclidean instruments alone. Notice how purely *intellectual* these problems are. (The restriction to Euclidean instruments of construction is equivalent to demanding that the construction can be *proved* successful by reasoning logically from Euclid's five postulates of geometry.) But each of these problems has spawned a wealth of mathematics by its very attractiveness and intractability – mathematics that has subsequently been found "useful", even indispensable, in the study of the physical world and the development of science and technology.

Here, in brief, are some other notable features in the history of mathematics:

1. The persistent phenomenon of *simultaneous and independent discovery* indicates that it is the community of mathematicians rather than isolated genius that takes major steps: the discoveries occur when the moment is in some sense "ripe"; the necessary ideas are "in the intellectual air".
2. *Solutions spawn more problems*: when a problem is solved, it is usual for many more problems to grow promptly in its place, out of its solution.
3. *Unexpected liaisons* between apparently distant fields of mathematics occur astonishingly frequently, with consequent fruitful *cross-fertilisation*. A solution to one problem may arise from an unexpected quarter, and then have consequences for another problem remote from both.
4. The course of development of any mathematical field displays a pattern of *recurrent diversification and reunification*. Results and problems may proliferate alarmingly, until, with some novel insight, they are wonderfully brought to a new unity as instances of a more

general theory. Solutions to problems in the new context can yield immediate solutions to many disparate problems in diverse contexts. Proofs become dramatically shorter and more elegant.

4. Mysterious reciprocities

Through constantly posing and solving problems (mathematicians say "conjecturing and proving theorems"), the mathematical community is engaged in constructing (Platonists will say discovering) a vast, unified, abstract edifice of breathtaking beauty, intricate design and ever-growing vitality. This landscape of the mind is both astonishingly attractive to the aesthetic sense of the mathematician (who regularly uses words like "elegant" and "beautiful"), and also strangely poised to grasp and express the nature of physical reality. Galileo famously preached that one should "sit before the facts" like a child, but at least as important as his empiricism was his recognition of mathematics as the key. God's "second book" of nature was written in the language of mathematics, and humans, created in God's image, were capable of learning to read that book. In the time of Galileo, Descartes and Kepler, the rational and the empirical, "head and hand", were coming together in a novel and fruitful union. Galileo, making his own telescope in his own workshop, was one of the first of a new breed of philosophers willing to get their hands dirty. Ever since, there has developed, between the world of pure mathematics and the physical universe, a fruitful interaction, experienced by the mathematical community as a kind of dialogue or conversation.

On the one hand, mathematics created and developed by human minds for its own sake turns out (often much later) to be the perfect vehicle to capture and express, almost miraculously, the structure of the universe. Space does not permit examples here, but some eminent witnesses can be called, each speaking in quasi-religious terms, and each invoking the idea of mystery:

> Nothing is more impressive than the fact that, as mathematics drew increasingly into the upper regions of ever greater extremes of abstract thought, it returned back to earth with a corresponding growth of importance for the analysis of concrete fact... The paradox is now fully established that the utmost abstractions are the true weapons with which to control our thought of concrete fact. (Whitehead 1926 p. 47)

> I have never found a better expression than the expression "religious" for this trust in the rational nature of reality and its peculiar accessibility to the human mind. Where this trust is lacking science degenerates into an

uninspired procedure. Let the devil care if the priests make capital out of this. There is no remedy for that. (Einstein 1951a)

Einstein elsewhere (1951b) describes this phenomenon of the comprehensibility of the world as "a miracle or an eternal mystery", and, lest his correspondent think that, weakened by age, he has fallen into the hands of priests at last, he adds: "Curiously, we have to be resigned to recognise the 'miracle' without having any legitimate way of getting further."

Eugene Wigner (1959) spoke movingly of "the unreasonable effectiveness of mathematics" as "a wonderful gift which we neither understand nor deserve". Bourbaki (1950 p.221), taking a detached view, observed that:

> from the axiomatic point of view, mathematics appears ... as a storehouse of abstract forms – the mathematical structures; and it so happens – without our knowing why – that certain aspects of empirical reality fit themselves into these forms as if through a kind of pre-adaptation.

For Paul Dirac (1963 p.53):

> It seems to be one of the fundamental features of nature that fundamental physical laws are described in terms of a mathematical theory of great beauty and power ... Why is nature constructed along these lines? ... We simply have to accept it.

He goes on to speak tentatively of God as mathematician. Sir James Jeans, earlier in the same period of dramatic, unexpected, fruitful liaisons between physics and pure mathematics, had also observed that God was beginning to appear as a pure mathematician. (God had been famously depicted by Plato as a geometer, and by Jacobi in the mid-nineteenth century as an arithmetician.)

On the other hand, the problems posed by grappling with the physical world, in the quest to understand its nature and master it, regularly give rise to new mathematics of great beauty. Dirac, of his fundamental work in quantum theory, felt moved to say:

> I think there is a moral to this story, namely that it is more important to have beauty in one's equations than to have them fit experiment.

He went on quickly to temper and justify this apparent scientific heresy:

> It seems that if one is working from the point of view of getting beauty in one's equations and if one has really a sound insight, one is on a sure line of progress. If there is not complete agreement between one's work and

experiment, one should not allow oneself to be too discouraged, because the discrepancy may well be due to minor features that are not properly taken into account and they will get cleared up with further developments of the theory.

Profound questions are raised by this spectacle of a community of minds in sustained, evolving dialogue with reality. Why do the abstract creations of minds describe nature so well? Why does the beautiful work, and why is what works so beautiful? There seem to be no limits to the complexity of the abstract structures mathematicians may freely create; but are there limits to the complexity of the universe? Will we run out of questions there? Will mathematics lose its vitality in abstraction for its own sake, divorced from the discipline of grounded "real" problems? Most scientists would say "No", with Haldane: "The universe is not only more mysterious than we think, but more mysterious than we can conceive..." The challenges will never cease. Are there (as Haldane may be taken to suggest) limits to the capacity of our minds? Will we run out of answers? Will it all get too messy, abstract and abstruse? Are there unfathomable depths, uncrossable gulfs, unclimbable cliffs? Are there conceptual dead-ends, analogous to those in biological evolution? Does Gödel's undecideability theorem prove the inadequacy of mathematics to provide answers to problems posable in its own terms? The lessons of history and the mood of the contemporary scientific and mathematical communities suggest a firm "No!" to all these questions.

For an insider or eavesdropper on this conversation, it feels something like a cosmic love affair. We are a match for nature, and reciprocally she is a match for us, as if a divine Matchmaker were behind the relationship. The reality with which we engage is both mysteriously accessible and profoundly perplexing; constantly inviting, sometimes rewarding, but often tantalisingly elusive, calling like a teasing lover from beyond the next rise or around the next corner. Sometimes she seems to come running to meet us, arms open wide. Sometimes there is a deathly silence and painful absence of response to our overtures. Despair threatens, but perseverance is always rewarded.

The overall mood of this conversation is positive and joyful. We do not perceive ourselves as extorting answers from a sullen enemy or victim; rather we are engaged in a dialogue – in the gentler art of questioning, of wooing something we love. And the answers we receive are sufficiently satisfying and provocative to challenge us to frame new questions. There is a richness about this conversation that engages all that is within us – all that is human – and calls for a fine balance of qualities. High creativity in any problem-solving arena seems to depend on a balanced alternation and tension between conscious and unconscious, discipline and relaxation, tradition and revolt, perseverance and serendipity. The best mathematics arises from interplay between logic and imagination, between rigour and intuition. Somewhat similar creative tensions occur in science (rational-empirical, theory-experiment), in religion (doctrine-experience, dogma-practice), in the arts (theme-variations, structure-improvisation), and in the mechanisms of natural selection (replication-mutation).

Richard Courant, speaking at Göttingen on the centenary of David Hilbert's birth, drew from the life and work of that great mathematician this lesson:

> Living mathematics rests upon a fluctuation between the antithetical powers of intuition and logic, the individuality of "grounded" problems and the generality of far-reaching abstractions. We ourselves must prevent the development being forced to only one pole of this life-giving antithesis.
> (Quoted in Reid 1970 p.220)

5. Conclusion: proposed basis-of-faith for the mathematical community

In conclusion, we give a set of tacit foundational beliefs that most mathematicians will assent to, and that constitute a good basis for the faith that unites and motivates the community. Given the extraordinary fruitfulness of this community, such a manifesto is pertinent to any study of mind and evolution, human psychology and creativity.

1. There is something out there, taking both concrete and abstract forms, of seductive beauty and great value, calling for exploration and study.
2. We can make sense of it and understand it; there do exist solutions to our problems, proofs or disproofs of our conjectures. There is congruence (consonance, compatibility, mutuality) between mind and reality. Eager questioning and hard work will be rewarded, and the unpredictable "Eureka" moment will surely come. Our understanding is unlimited in principle. We are a match for reality.
3. The life-blood of mathematics is great problems, both the purely intellectual and those generated by interactions with the physical world.
4. Our theorems, our concepts and our logical frameworks are *true*, or *correct*, by internal mathematical criteria such as freedom from contradiction; hence mathematics is perhaps as cumulative a discipline as human knowledge allows. Our representations (our *modelling*) of external reality, however, are always provisional.

5. Mathematical results, proofs and theories are valued by a complex set of criteria that include:
   (i) elegance or beauty, assessed by something similar to aesthetics;
   (ii) depth, assessed by logical distance from axioms;
   (iii) internal or mathematical fruitfulness, assessed by the multiplicity of connections with other results or fields, by generality (previous results subsumed) and by potency (open problems illuminated and new results deduced);
   (iv) external fruitfulness, assessed by actual or potential applications in the natural or social sciences.
6. Pure and applied mathematics are not to be separated, and all of mathematics should be encouraged and funded, not just the parts perceived to be "useful" in the short term. Beautiful mathematics is its own justification, but it is certain to find application eventually.
7. Mathematical truths (results and their proofs) are communally attained, ratified and warranted. Careful citation and giving others their due is considered far more important than mere priority.
8. We seem doomed, as a community and as individuals, to oscillate perpetually between the two poles of constructivism and Platonism – the consciousness of *creating* mathematics, and the sense of exploring or *discovering* the concepts and truths of mathematics. Perhaps these two apparently opposing views can be brought into some consonance by using the old image of Adam naming the animals: our "names" are indeed our creation, while the reality we seek to name is really "out there" independently of us.
9. Our making of mathematics, together with our progressive naming of the reality out there, is an end in itself, meriting total commitment, eliciting the utmost passion, and giving incomparable joy.

References

Bourbaki, N. 1950. *American Math. Monthly* 57, 221.

Dirac, P.A. M. 1963. The evolution of the physicists's picture of Nature. *Scientific American* 208, 53.

Einstein, A. 1951a. From a letter to Maurice Solovine, quoted in Jaki, S. *Cosmos and Creator*. Edinburgh: Scottish Academic Press, 1980, 52.

—. 1951b. From another letter to Maurice Solovine, quoted in Jaki, *ibid.*, 53.

Kline, M. 1972. *Mathematical Thought from Ancient to Modern Times*. New York: Oxford University Press.

Reid, C. 1970. *Hilbert*. New York: Springer-Verlag.

Whitehead, A.N. 1926. *Science and the Modern World.* Cambridge: University Press.

Wigner, E. 1959. Richard Courant Lecture, New York University; published in *Communications in Pure and Applied Mathematics,* 13.1, Feb. 1960.

Editor's Note

Readers interested to pursue the concept of mathematics as constructed by the human mind may like also to consult:

Lakoff, G. & Núñez, R.E. 2000. *Where Mathematics Comes From: How the Embodied Mind Brings Mathematics into Being.* New York: Basic Books

CHAPTER FOURTEEN

ARE ATTITUDES TO LIFE CHANGING? – THE EMERGENCE OF NEW MORAL INTUITIONS

R.I. VANE-WRIGHT

Dr Richard ("Dick") Vane-Wright (Hon DSc, Copenhagen) read Zoology at University College, London, and spent his professional life in the Dept of Entomology, the Natural History Museum, retiring as its Head in 2004. He has since held a NESTA (National Endowment for Science, Technology and the Arts) Fellowship at the Durrell Institute of Conservation and Ecology, University of Kent, and is now Honorary Professor of Taxonomy there. Books that he has co-authored or co-edited include "The Biology of Butterflies" (1984), "Systematics and Conservation Evaluation" (1994), and "The Seymer Legacy" (2005).

Here he considers the wide range of responses to the present ecological challenge, and suggests that the changes required in philosophical and theological standpoints, if humanity as we know it is to survive, will constitute a further step in the evolution of mind.

Opening Quotations

Why should we care about the Earth when our duty is to
the poor and sick among us? God will take care of the Earth.
—*Attributed to Mother Teresa by James Lovelock (2006 p2)*

The planet takes care of us, not we of it.
—*Lynn Margulis (1999 p143)*

The choice is ours: form a global partnership to care for Earth and one another, or risk the destruction of ourselves and the diversity of life … We must realize that when basic needs have been met, human development is primarily about being more, not having more.
—*Earth Charter International Council (2000)*

Introduction

I have been concerned with biological diversity since 1980, after witnessing unthinking destruction of the central Philippine rainforests. Progressing from area selection technology to governance issues, I have now switched to a cultural approach (cf. Callicott 1994): studying different worldviews and the attitudes they engender to the natural world in general, and biological diversity in particular. My aspiration is to encourage all people, consistent with their own beliefs, to recognise their identity with the rest of nature (Naess 1985), and make conservation of other living things an imperative rather than a reluctant afterthought. This has led me to wonder how knowledge based on faith and knowledge based on science interact (Passmore 1985 p141) in fashioning our intuitive morality – that sense of what is right and wrong regardless of rational arguments. How do we *feel* about the natural world, about otherness, about diversity – and can those feelings change?

The "greening-of-religions phenomenon" suggests that religions themselves are being transformed as a result of growing human awareness that planetary systems are finite – to the point where "a kind of civic planetary earth religion may be evolving" (Taylor 2004; cf. Earth Charter International Council 2000). This may be due to the impact of scientific thinking: "If it weren't for ecology we would not be aware that we have an 'ecologic crisis'" (Callicott 2005). Even so, Callicott (1994) suggested earlier that religious and other spiritual disciplines have much to offer in our environmental predicament – as advocated in different ways by Nasr (1976), Berry. W. (1982) and Rolston (1999), and numerous other commentators on the relationship between humanity and the rest of nature. First, however, I would like to consider three major scientific paradigm shifts that have affected our understanding of "nature" and our place within it. This three-fold revolution, occurring over the past 500 years, appears at least partly responsible for changes in religious interpretation, perception, and even preoccupation.

Copernicus and Cosmology

The realisation that the Earth is not at the centre of the Universe, but is just one of several planets revolving around an average star in an average galaxy in a universe so large, complex and old as to defy imagination, was ushered in by the work of Nicolaus Copernicus. Nearly 500 years later we are still entranced by this emerging vision, trying to draw fundamental understanding from glimmers of light that started out from distant galaxies countless millions of years ago. To look into deep space is to connect with unbelievably remote events – in distance and in time.

Darwin and Organic Evolution

Life has been around for a long time too – at least three billions years – but it is still only known to us from Earth. Since its origin or introduction, there have been massive changes to the planet, and to life itself. After two billion years of microbial evolution, including concomitant changes in the atmosphere, multi-cellular animals appeared in the seas about 650 million years ago (Margulis 1999). Colonisation of dry land followed about 150 million years later, heralding land plants, insects, mammals and birds.

Although the great majority of species that ever lived have long gone extinct, when *Homo* evolved a mere 5–6 million years ago, life on Earth was probably approaching an all-time biodiversity maximum (Rohde & Muller 2005). Humanity emerged into a world occupied by as many as 15 million or even more kinds of other living organisms (accurate assessments remain very uncertain). Although we are now at the beginning of a human-induced mass extinction, most of those species are still with us. The consequences of their wholesale loss are unknown.

Two huge developments in our understanding of life have brought us to a second and now a third "Copernican revolution". The first started with Linnaeus and other 18th century taxonomists, who paved the way for the Darwin/Wallace 19th century theory of evolution by natural selection. As a result, we were not just relegated from the centre of the universe, but our oft-assumed separateness from the rest of nature was no longer assured. Humans might only be quantitatively, not qualitatively different from other living things. This shift in awareness probably remains the greatest challenge to theology.

Lovelock and Earth Systems Science

The third key change started in the 1920s, with the pioneering research of population ecologist Charles Elton and systems theorist Ludwig von Bertalanffy, and numerous others. This led to our current scientific understanding of the nature of life.

Living things are self-organising and autonomous, or *autopoietic* unities (Maturana & Varela 1998 p47). All living things are ecosystem inter-dependent, being organised in networks, within nested systems, reliant on flows of energy and materials, subject to cycles and undergoing development, and maintained in dynamic organization far from thermodynamic equilibrium by cybernetic feedback mechanisms (Capra 2005). Third, all living things now found on this planet appear interconnected back through time in an uninterrupted stream (the standard genetic code used by the vast majority of organisms being the best evidence for this – although there are other explanations for this observed pattern: Knight *et al.* 1999). These insights apply as much to *Homo sapiens* as they do to the millions of other living species on Earth, and the even greater numbers of species thought to have existed in the geological past.

These ideas, that living organisms are autopoietic, thermodynamic systems operating far from equilibrium and embedded within networks, and always interconnected through complex positive and negative feedback loops, came together dramatically in the 1970s as James Lovelock's *Gaia Hypothesis* (Lovelock 1979). The Earth's biosphere acts as a self-regulating, homeostatic system, involving the interrelationships of both living and non-living elements, favourable to the continuing existence of life itself (Capra 1996; Harding 2006).

Earth Systems Science now encompasses the unfolding totality of life on this planet, and the complex relationships between its myriad components, living and non-living. In reaching this understanding, new forms of mathematics had to be developed to deal with the complex, non-linear systems that give rise to emergent properties and chaotic events. These systems have maintained a shifting, dynamic regime within which life, on our finite Earth, has survived through an unending evolutionary process over billions of years (Wilson 2002 p39). The fresh understanding that general systems theory, and the *Gaia Hypothesis* in particular, have provided is that the totality of interactions within the biosphere is responsible for its fitness for human life, and that the interactions of all things within the biosphere are consequential – although not in any simple, direct cause and effect way (Macy 1991 p170).

Thus the causal chains affecting the biosphere are complex and highly contingent. Even so, the mass effect of millions of seemingly trivial individual actions, such as chopping down one tree or burning a single lump of coal, do add up. Human appropriation of net primary productivity already exceeds 25% in support of our single species (Haberl *et al.* 2007). This, and our release into the atmosphere, over just a few decades, of vast quantities of fossil carbon that took the biosphere millions of years to sequester, now means that the collective impact of six billion people threatens to destabilise the entire planetary system on which we, and most familiar forms of multi-cellular life, are dependent.

Even 50 years ago only a few visionaries could imagine the consequences of our collective actions being so great – the "forces of nature" seemed overwhelming compared to our puny efforts. While nature may yet overpower us (Lovelock 2006), there can be no doubt that we will be held responsible by our descendants for whatever happens as a result of our disregard of nature – not because of a judgemental godhead, but because we now understand how life on Earth really works. We know what we are doing – how much longer can we afford to be in denial?

Worldviews

Current worldviews have developed at different historical periods with respect to our collective understanding of humanity, the cosmos, biological evolution, and our place in nature. In particular, all the major religions predate post-Copernican scientific understanding (Smart 1998 pp30–31). However, regardless of our ability to understand physics, chemistry, cybernetics and biology, the cumulative shifts in perspective, from humanity being set apart from the world to inextricably being part of it, appear to herald a transformation in the way in which people *feel* about nature and our place within it. Is there evidence for such a claim?

Following publication of *General System Theory* (von Bertalanffy 1968), numerous books have appeared that, although academic in nature, seek a wide audience to share the insights of the three-fold scientific revolution. One of the first and still among the most significant is *Steps to an Ecology of Mind* (Bateson 1972). Others include Berry (1988), Macy (1991), Roszak *et al.* (1995), Capra (1996), Berry (1999), Harding (2006), and Lovelock (2006).

In my view, these books – and many others like them – have established a 21st century Zeitgeist, as the following quotations reflect:

> We are aware that the earth was born and is borne by a delicate dynamic of forces … [and] are enchanted by the splendour of its life emergence, of which we are but a part" (Bassett *et al.* 2000 p7).

> To grasp this change [in understanding] calls for an unparalleled upheaval in our moral consciousness (Midgley 2004 p171).

> The green shift seems to require a root change of human outlook, a mutation of collective philosophy, a spiritual phase transition (Kearns & Keller 2007 p.xi).

> Religions demonstrate that they can change, transforming themselves in response to new ideas and circumstances (Tucker & Grim 2007).

At the heart of this report is the belief that the world is not just changing but *transforming* ...to one in which both science and spirituality reconfigure our most basic understandings of human consciousness and how to live harmoniously in a healthy and sustainable ecosphere" (O'Dea 2007).

Changing Reality?

James O'Dea's words relate to a fundamental tension in metaphysics: is reality "out there", to be discovered, or is it a product of the conscious human mind? In moving from the *Tractatus* to *Philosophical Investigations*, Ludwig Wittgenstein "abandoned the idea that the structure of reality determines the structure of language", and replaced it by the notion that "our language determines our view of reality" (Pears 1971 p13). If so, "language" cannot be fixed, as our view of reality passes, from time to time, through revolutions (Kuhn 1970) of the sort wrought by Copernicus, Darwin and Lovelock, and perceived by O'Dea.

Changing Morality?

Are moral norms likewise subject to revolutions, as we assimilate or create new understandings of reality? As we now appreciate, as never before, our place in the universe, our oneness with the rest of life, and our interdependence with it, are human attitudes to life changing? As we absorb the implications of Copernicus, Darwin and Lovelock, are we witnessing the emergence of new moral intuitions regarding our attitudes to life? Are we pushing back moral consideration from humans and domesticated animals to all other forms of life, even the entire planet – as Arne Naess (e.g. 1986) and the deep ecologists have long urged us to do?

The general question "What values should we live by" (Fox 2006 p53) transforms within an integrated worldview to the practical issue "How are we to act ... in this world?" (Aerts *et al.* 1994/2007). In Heylighen's (2000) cybernetic model of a worldview, action is the result of the interaction of our current knowledge and understanding of our environment with our internal values, morals and goals. This is similar to Piaget's model of cyclical cognitive assimilation and accommodation within the context of individual action schemata (Piaget 1971). Or, as explored by Maturana & Varela (1998), "cognition involves perception, emotion and action" (Pretty 2002 p151). To me this implies that internalised values, goals and even our moral sense can be altered through successive assimilation cycles, as our model of reality and our relationship to reality itself changes and develops through experience. As Passmore

(1995 p141) has put it, "The emergence of new moral attitudes to nature is bound up ... with the emergence of a more realistic philosophy of nature."

Such a view would be rejected by those who see morality as fixed by the Creator (or by "divine commands": Norman 2004 p87), and not subject to revision by evolution of the human mind – individual or collective. The idea that morality is fixed, and has essential, eternal attributes, is embraced for example by Seyyed Hossein Nasr. Nasr is not only opposed to humanism, but also opposes the theory of organic evolution: "A species could not evolve into another because each species is an independent reality [that is] qualitatively different ..." (Nasr 1976 p124), a deep conviction that he still holds today. Appeals to essentialism have been rejected by most philosophers concerned with organic evolution – although a striking case for a form of resurrection has recently been made by Walsh (2006).

The evolving mind – unfinished business?

W.B. Yeats prefaced his poem *Responsibilities* with the epigraph "In dreams begins responsibility." If dreams beget responsibility, does understanding beget morality? "For one species to mourn the death of another is a new thing under the sun" (Leopold 1949 p117). That all forms of life have intrinsic worth is now seen as a new value, one that requires development of an integrated worldview that can offer "a reasoned framework for environmental action and feeling" (Aerts *et al.* 1994/2007 p34). If so, I suggest this represents nothing less than the emergence of a new moral intuition, based in part on the three revolutions in scientific understanding represented by cosmology, organic evolution, and systems theory.

The evolution of the collective mind with respect to what is right and wrong must have theological significance. This implies that those engaged in the greening of religion have a potentially difficult choice. If, theologically, they wish to embrace morality as an unchanging essence, they may have to deny Leopold's claim. On the other hand, if they embrace Leopold's view, then they have to deal with the difficulty of not just the *origin* of the human mind in an evolutionary sense, but with its *ongoing* development, arguably deeply affected by science as one element of the fourfold wisdom (indigenous, female, traditional, scientific: Berry, 1999). According to Thomas Berry, indigenous wisdom concerns intimacy and participation in the functioning of the natural world, female wisdom concerns the need to balance androcentrism with nurture and self-transcendence, traditional wisdom expresses the need to embrace the spiritual realm, and scientific wisdom concerns the realisation that the universe has come into existence by a process of unending evolution and emergence over an immensely long period of time. In short, those desirous of taking the greening of religion forward will need to develop what amounts to an ecologically sensitive theology (Boff 1995 p43). In this context, does the notion of evolution of the human mind (Laszlo 1987 p116) as unfinished, as work still in progress, represent a special challenge for theology?

References

Aerts, D., Apostel, L., De Moor, B. *et al.* 1994. *Worldviews: From fragmentation to integration.* Brussels: VUB Press. Internet edition (Vidal, C & Riegler, A. 2007) *http://www.vub.ac.be/CLEA/pub/books/ worldviews.pdf* (accessed 22 Jan 2008)

Bassett, L., Brinkman, J.T. & Pedersen, K.P. 2000. *Earth and Faith. A book of reflection for action.* New York: Interfaith Partnership for the Environment, and United Nations Environment Programme.

Bateson, G. 1972. *Steps to An Ecology of Mind.* New York: Chandler.

Berry, T. 1988. *The Dream of the Earth.* San Francisco: Sierra Club.

—. 1999. *The Great Work. Our way into the future.* New York: Random House.

Berry, W. 1982. *The Gift of Good Land. Further essays cultural and agricultural.* New York: North Point Press (Farrar, Straus & Giroux).

Boff, L. 1995. *Ecology and Liberation. A new paradigm.* Maryknoll, NY: Orbis.

Callicott, J.B. 1994. *Earth's Insights.* Berkeley: University of California Press.

—. 2005. Natural history as natural religion, in *Encyclopedia of Religion and Nature,* B. Taylor & J. Kaplan (eds.) 1164–1169. London: Continuum.

Capra, F. 1996. *The Web of Life.* New York: Doubleday.

—. 2005. Speaking nature's language: principles for sustainability, in *Ecological Literacy*, M.K. Stone & Z. Barlow (eds.), 18–29. San Francisco: Sierra Club.

Earth Charter International Council. 2000. *The Earth Charter.* *http://www.earthcharterinaction.org/2000/10/the_earth_charter.html* (accessed 22 Jan 2008).

Fox, W. 2006. *A Theory of General Ethics. Human relationships, nature, and the built environment.* Cambridge, MA: MIT Press.

Haberl, H., Erb, K.H., Krausmann *et al.* 2007. Quantifying and mapping the human appropriation of net primary production in Earth's terrestrial ecosystems. *Proceedings of the National Academy of Sciences* 104, 12942–12947.
Harding, S. 2006. *Animate Earth.* Dartington, Devon: Green Books.
Harman, W. 1998. *Global Mind Change.* San Francisco: Berrett-Koehler.
Heylighen, F. 2000. What is a world view? *Principia Cybernetic Project. http://pespmc1.vub.ac.be/WORLVIEW.html* (accessed 22 Jan, 2008)..
Kearns, L. & Keller, C. 2007. *Ecospirit. Religions and philosophies for the Earth.* New York: Fordham University Press.
Knight, R.D., Freeland, S.J. & Landweber, L.F. 1999. Selection, history and chemistry: the three faces of the genetic code. *Trends in Biochemical Sciences* 24, 241–247.
Kuhn, T.S. 1970. *The Structure of Scientific Revolutions* (2nd edn). Chicago: University Press.
Laszlo, E. 1987. *Evolution. The Grand Synthesis.* Boston & London: Shambhala.
Leopold, A. 1949. *A Sand County Almanac.* New York: Oxford University Press.
Lovelock, J.E. 1979. *Gaia.* Oxford: University Press.
—. 2006. *The Revenge of Gaia. Why the Earth is fighting back – and how we can still save humanity.* London: Penguin.
Macy, J. 1991 *Mutual Causality in Buddhism and General Systems Theory.* Albany: State University of New York Press.
Margulis, L. 1999. *The Symbiotic Planet. A new look at evolution.* London: Phoenix (Orion Books).
Maturana, H.R. & Varela, F.J. 1998 [1987]. *The Tree of Knowledge. The biological roots of human understanding* (revised edn). Boston: Shambhala.
Midgley, M. 2004. *The Myths We Live By* (pbk edn). London: Routledge.
Naess, A. 1985. "Identification as a source of deep ecological attitudes." In *Deep Ecology* (ed. M. Tobias) 256–270. San Diego: Avant Books.
—. 1986. Intrinsic value: will the defenders of Nature please rise? in M.E. Soulé (ed.)*Conservation Biology*, 504–515. Sunderland, MA: Sinauer.
Nasr, S.H. 1976. *Man and Nature. The spiritual crisis of modern man* (Mandala edn). London, Allen & Unwin.
Norman, R. 2004. *On Humanism.* London: Routledge.
O'Dea, J. 2007. *Evidence of a World Transforming.* Petaluma, CA: Institute of Noetic Sciences.
Passmore, J. 1995. Attitudes to nature, in *Environmental Ethics,* R. Elliot (ed) 129–141. Oxford: University Press.

Pears, D. 1971. *Wittgenstein*. London: Fontana/Collins.
Piaget, J. 1971. *Biology and Knowledge. An essay on the relations between organic regulations and cognitive processes*. Trans Walsh, B.. Edinburgh: University Press.
Pretty, J. 2002. *Agri-Culture. Reconnecting people, land and nature.* London: Earthscan.
Rohde, R.A. & Muller, R.A.. 2005. Cycles in fossil diversity. *Nature* 434, 208–210.
Rolston III, H. 1999. *Genes, Genesis and God.* Cambridge: University Press.
Roszak, T, Gomes, M.E, & Kanner, A.D. (eds) 1995. *Ecopsychology*. San Francisco: Sierra Club.
Smart, N. 1998. *The World's Religions* (2nd edn). Cambridge: University Press.
Stewart, I. 1998. *Life's Other Secret. The new mathematics of the living world.* London: Penguin.
Stone, M.K. & Barlow, Z. (eds) 2005. *Ecological Literacy*. San Francisco: Sierra Club.
Taylor, B. 2004. A green future for religion? *Futures* 36, 991–1008.
Tucker, M.E. & Grim,J. 2007. Daring to dream: religion and the future of the Earth. *Reflections*, Yale Divinity School (Spring issue) 4–9.
von Bertalanffy, L. 1968. *General System Theory*. New York: Braziller.
Walsh, D. 2006. Evolutionary essentialism. *The British Journal for the Philosophy of Science* 57, 425–448.
Wilson, E.O. 2002. *The Future of Life*. New York: Knopf.

Acknowledgements

I thank the Science & Religion Forum for the invitation to attend the original conference and submit my presentation for publication. Professor Richard Norman and Dr David Lorimer kindly read the MS; however, any errors or misunderstandings that remain are entirely my responsibility. Rev. Dr. K.C. Abraham introduced me to the work of Leonardo Boff. I also thank NESTA for support from 2005-2008.

CHAPTER FIFTEEN

NOLITIONAL FREEDOM AND THE NEUROBIOLOGY OF SIN

RON CHOONG

Revd Dr Ron Choong read law, science, international relations, the humanities and theology during the 1990s, in both Britain (London and Open universities) and the United States. He was ordained into the Presbyterian Church in America in 1999 and the following year began studying simultaneously for an STM at Yale and a PhD at Princeton Theological Seminary – the latter under the direction of Prof. Wentzel van Huyssteen. In 2003, he founded ACT – a research and training resource center for Christian apologetics with reference to history, philosophy, science, and world religions. He also conducts a summer preaching and teaching ministry in Europe, Asia, and Africa, and has recently (2008) published "Thinking Things Through in Theological Safe Space".

Ron Choong's research interests include the emergence of intelligence and evolution of moral cognition as well as the neurotheology of human consciousness. Here he brings all these together in addressing perhaps the toughest problem at the interface of science and theology, the moral responsibility of a biologically-selected and electrochemically operating brain/mind.

The essential paradox

Are we free not to sin? In both popular imagination and theological doctrine, freedom of will and sin are intertwined. However, the received wisdom of the 21st C is that we are born as doomed and determined sinners, but punished as agents of free will who consciously choose to reject God's grace – biologically determined to sin yet theologically punished for it. This paradox has never been satisfactorily accounted for.

However, with advances in neurobiology, it has come under scrutiny. Can studies into how our behaviour may respond to electrochemical changes in our brains explain, or explain away, why we do what we are sure we ought not to do? Does the electro-chemistry of brain functions determine our behaviour and hence our soteriological[1] status?

Guilt in theology and law

If we are deemed to be sinful because we commit sins, that implies that when we do so we are guilty before God. On the other hand, if we sin because we are made sinful, we are merely living out our biologically created destiny.

Again if, as humans, we were made to be different from animals, i.e., *to possess a moral grammar and to respond to the divine order,* are we also given the capacity of performance to meet this divine demand, and the judgement to know the difference between it and our biological makeup? If not, ought we to be deemed guilty? The practical implications of this issue range from criminal justice to jurisprudence, from doctrinal theology to philosophy of mind, from treatment for psychological distress to matters of public health, and of course the doctrine of salvation itself. Central to this doctrine is the idea of guilt.

The notion of guilt is one of the pillars of modern societies because the justification of punishment relies on the existence of guilt. The effective operation of justice is obviously crucial in the successful development of societies because it elicits trust in the social institutions of government and rule of law; only thereby do people feel they can sleep in peace, safe in the knowledge that justice protects the innocent, i.e., the non-guilty. If no one is guilty, no one can be punished and the innocent cannot expect justice. So society's quick answer to the question above may be that *we are sinful because we sin*. This formulation entails the possibility of not sinning, and directly challenges the view that all humans are born in sin, i.e., doomed to exercise their propensity to sin and therefore, in need of salvation through divine forgiveness. In this first view, some of us can die never having actualized our potential for sin, so that while we are born with the capacity to sin, we never become sinners: examples would be when we die before we are capable of resisting God's grace (infant deaths) or we are unable to retain conscious resistance to that grace (mentally challenged from birth).

Many traditional Christian doctrines of atonement deny any human capacity, *per se*, to accomplish the goal of not having ever sinned. This

---

[1] Soteriology = the doctrine of salvation. Ed.

second view, that *we sin because we are innately sinful*, implies that even though we are morally conscious beings, we do not enjoy the capacity to behave morally all the time, and hence, are in need of salvation. Some have taught that we are morally deficient because we are spiritually immature. Only when we will to believe God and receive forgiveness may we be deemed righteous. But is this responsibility to believe an act of free will or are we evolved to such a volitional act?

In contrast, some contemporary doctrines of the atonement suggest that we evolved what moral cognition we possess, and cannot be blamed for our imperfect behavior since we are physiologically inclined to seek our biological survival, and this will often be to the detriment of others. Under this rubric, it is possible to hope that our evolving moral awareness will continue to increase and we shall one day meet the divine command for justice because God has made us capable of reaching the requisite capacity.

Fraser Watts' Christology (Chapter Four of this book) is an example of such an outlook. It relies on a Neo-Darwinian evolution of human consciousness in which Christ also influences the human power of increasingly good information-processing. This account seeks to avoid the teleological language of theology. But does such an evolutionary model, applied to Christ himself, survive theological scrutiny? Perhaps Watts refers to the works rather than the ontology of the incarnated Logos. If we take Christ to mean the incarnated second person of the divine order, then the evolution of Christ cannot violate the sanctity of an eternal member of the triune Godhead. This view of the evolution of Christological soteriology posits the Logos as having perfect ontological awareness but kenotically[2] experiencing an evolving epistemic[3] consciousness.

What is at stake here? Apart from the sociological, legal and cultural implications, there also exists an important theological matter. This involves both the affirmation of Jesus' historic resurrection and the hermeneutical[4] approach to the texts. Did the writers of the Gospel record historical and chronological events as they perceived them or did they take poetic license in weaving a theological account? Richard Bauckham (2006) warns us against abandoning the historicity of the gospels. He is right to offer such caution, yet himself must rely on selective records to make his case. The reality is that we cannot know with certainty what happened. We adopt the authorities of testimonial witnesses of the past and either receive or reject the accounts with various degrees of trust. I am

---

[2] Kenosis = self-limiting of divine power and attributes in incarnation
[3] Epistemic = knowing/understanding
[4] Hermeneutical = expository, interpretative. (All footnotes by Ed.)

indebted to both Steven Mithen (Chapter One) and Fraser Watts for their defenses of the *metaphorical priority of compositional language* (or what I prefer to call speech). A metaphorical reading of the Bible may be the primary code of linguistic intelligence but, more importantly, metaphor both escapes the rigidity of contextual meanings and transcends the boundaries of cultural expression. Metaphor in fact contains a precision that literality cannot sustain. The biblical writers were not writing chronological accounts but rather *kairological* accounts – describing historical events as having theological import.

How can we make rational sense of the scriptural texts? With advances in neurobiology, another resource of rationality emerges. But with what measure of assurance may we subject the Gospels to neurobiological inferences? Theological reflection demands a measure of fidelity to divine revelation that science cannot and need not meet. A discipline of inquiry that advances with new tentative conclusions replacing older tentative conclusions bears a different burden from a discipline of inquiry that takes its cues from the received texts of adopted authority. Any attempt to draw from both resources to assess interdisciplinary reasoning strategies must note the characteristics of these distinctive methods, which offer qualitatively differing sources of knowledge for understanding. With this caveat, let us consider what hints can be had from neurophysiology concerning the relationship between freedom, consciousness and sin.

Libet's experiment and its implications

What we think brain science can tell us depends on what we already believe about the fabric of reality. This does not mean that scientific investigation cannot advance our self-understanding, but that any hasty interpretation of the neurobiology of sin is fraught with the danger of bias and prejudice. A quest for convergence between science and theology offers a more robust prospect of understanding when undertaken under the critique of philosophy. The scientific aspect is reductionist but testable/falsifiable while the theological aspect is holistic but needs to meet the test of coherence. While each field of inquiry, on its own, need not submit to the tests of the other, a true interdisciplinary engagement demands a disciplinary generosity which makes even core beliefs vulnerable and accountable.

So, is consciousness merely a side-effect of neuronal functions – an epiphenomenon of brain states? I think not. But we have difficulty in even defining consciousness, and this is a clue to the complexity of awareness. Freedom of will demands awareness of our volitional actions, and there is a presumption of immediate cognitive awareness in Christian theological reflection. However, full mental awareness is obviously not a universal trait. Furthermore, focus on volition has undermined the importance of *cognitive nolition*, the capacity to veto a will to act. In the broad sense it is true that freedom demands awareness of our volitional actions. However, any capacity for judgement must involve nolitional as well as volitional freedom. The idea behind nolitional freedom is drawn from the work of Benjamin Libet. This becomes important as we consider the scientific aspect of sin and how we may move towards an approximate measure of what constitutes, for lack of a better term, the threshold for the "readiness potential" of sinful behavior.

In several experiments on consciousness which continue to baffle many commentators, Benjamin Libet (1985) registered a time delay of about 500 milliseconds between a measured mental activity and the conscious awareness of that activity[5]. This suggests that unconscious electrical processes in the brain ("readiness potentials") precede conscious decisions to perform some volitional, spontaneous acts. These experiments, performed more than 20 years ago, have not to date been successfully falsified. Their simplistic interpretation is that unconscious neuronal processes precede, and potentially cause, volitional acts which are retrospectively felt to have been consciously motivated by the subject. Neurophysiology thus suggests that our sense of conscious instigation for our acts is an illusion of retrospection. But one question is: does priority entail causation? Does the chronological priority of the readiness potential indicate its causal role for the posterior conscious awareness? Typical treatments in scientific studies tend to explain moral choices as sociobiological survival strategies and vestiges of adaptive advantages. This is in part due to the methodological naturalism that scientific investigation entails. But in an interdisciplinary dialogue where methodological naturalism is temporarily suspended, how does Libet's experiment on consciousness inform theological reflection on sin?

The theological concept of sin defines it as the conscious rejection of God's grace. Does Libet's proposal that we are never conscious of the present; that our visual perception is itself illusory so that we do not see what we are currently seeing; and that we never experience or know the "now", threaten this biblical/theological account of moral cognition, which seem to require immediate conscious volitional awareness? Do the biological accounts of how the brain works strain current theological interpretations of the story of a historic "Fall" in Genesis? The

---

[5] See also Introduction, and Chapter Seventeen (Ed)

Augustinian posit of an acquired and transferable original sin in a Fall event continues to dominate contemporary thought. Current biological, psychological and philosophical accounts of how the brain works all strain this conservative literalistic interpretation of the Genesis account. An analysis of how the doctrine of *imago Dei* was understood through the ages provides a possible solution to the impasse (Middleton, 2005). The phrase "falling upwards" embodies a more optimistic understanding of the Fall metaphor. This recovers an ancient interpretation which understood the Fall metaphorically and, indeed, describe it as a necessary event to culminate in moral maturity. Could moral cognition (like an inhibitor in synaptic transmission!) operate not only as a cognitive sensor guiding us *on what we will* in our choices but *also on what we nill*, i.e., on restraining our exercise of freedom of choice? *Negation of will follows prior freedom to will*, i.e., pre-conscious volition precedes conscious nolition but on can be vetoed by it. The analogy should not be lost to any biblical scholar who notes the divine gift of freedom in the Garden of Eden followed by the qualifier that Adam and Eve should nill any desire to eat of the fruit of knowledge of good and evil.

Benjamin Libet himself believed that while we cannot be responsible for our unconscious urges, we can certainly be responsible for our consciously controlled choices. Perhaps moral cognition and the divine call to holiness involve the choice to nill as well as to will. It appears that neurobiology has not done away with the notions of sin and guilt because freedom of choice remains intact despite Libet's *prima-facie* challenging observations.

I propose a model in which we acknowledge the existence of preconscious volition but note the conscious power of veto permitted by the time-delay of consciousness. This preserves the theological demand for personal guilt while also accounting for the evolutionary emergence of moral cognition as the necessary result of increasing sentience. Central to this argument is the claim that human consciousness is subjectively irreducible.

Is there then a neurobiology of sin and does it explain the arrow of causality? I think a reflective theological account must maintain the primacy of sin and guilt because we enjoy the power of cognitive veto. Therefore, we are sinful because we sin. I conclude that the Christian doctrine of sin remains a sustainable teaching – that as creatures with a modicum of free will and the capacity for moral discourse, we bear responsibility for our nolitional volitions.

References

Bauckham, R. 2006. *Jesus and the Eyewitnesses: The Gospels as Eyewitness Testimony*. Grand Rapids: Eerdmans.

Libet, B. 1985. "Unconscious cerebral initiative and the role of conscious will in voluntary action", *Behavioral and Brain Sciences* 8, 529-566.

Middleton, R.J. 2005. *The Liberating Image: The imago Dei in Genesis 1*. Grand Rapids: Brazos Press.

Chapter Sixteen

The Healing Instinct: Functionality, Integrity and Relationship – Holistic Principles in Evolution

Jeremy Swayne

Revd Dr Jeremy Swayne served as a medical doctor for 40 years, with parallel qualification as a homeopathic practitioner. Towards the end of his professional life, he was also ordained into the Church of England ministry.

In this short paper, printed exactly as contributed to the 2007 conference, he proposes that an instinct for wholeness, expressed in the experience of healing, realizes its full potential when the ordinary human being achieves convergence with Christ Jesus.

Introduction

If you know the Flanders and Swann song about the honeysuckle and the bindweed, you will recall that their marriage was thwarted by their parents' objections to the fact that the two families spiralled in opposite directions towards the sun, one clockwise and the other anticlockwise. Ultimately the two star-crossed lovers were found to have pulled up their roots and withered away. Science and religion follow similarly different but complementary trajectories towards the light, and will similarly fail to thrive if their marriage is denied. If the honeysuckle and the bindweed had intertwined with the intimacy they desired, they would have formed the intertwined pattern of the Caduceus, incorporating the staff of Aesculapius and a symbol of healing – to my mind the most appropriate image for this relationship[1].

I have used this cottage garden metaphor, for the sake of the healing symbolism that it introduces, and also, in a sense, to bring Science and Religion down to earth. I am not an academic, but I greatly appreciate the academic emphasis in the work of the Science and Religion Forum, for the stimulus and the conceptual framework it has provided for my own thinking. My acquaintance with the Forum is recent and I am unlikely now to be able to do more than dip my toe in the rich ocean of thought and literature to which it has introduced me. But my enthusiasm for its work and ideals arises from the recurring challenge of reconciling concepts of science and religion, more specifically medical science and Christianity, in the everyday lives of ordinary folk, particularly my patients and, since my ordination of seven years ago, parishioners. The common ground for this has been the search for healing and the experience of healing, and the exploration of human wholeness that is the goal of the healing process.

It is this experience that provides the perspective on theology, evolution and the mind that I am asking you to consider. That formal conference title represents for me a dynamic evolutionary process in individual people's lives; a dynamic that is, in the words of John Hedley Brooke's recent appreciation of Arthur Peacocke, "replete with potentiality and the possibility that humans might develop into more Christ-like persons" (Brooke 2006).

Propositions

This paper offers the following propositions:
i. That evolution reached its climax with the birth of Christ.
ii. That this is reflected in a conscious understanding of the divine destiny of humans, and indeed of creation.
iii. That this understanding and the whole thrust of evolution is holistic, both in terms of individual human wholeness and the eco-system.
iv. That this holistic evolutionary dynamic has three components – functionality, integrity and relationship.
v. That within this dynamic healing and health maintenance processes are not only an evolutionary imperative but also essential to its creative momentum.

---

[1] Anyone unfamiliar with this symbol can find a drawing and description at <http://en.wikipedia.org/wiki/Image:Caduceus.svg> Ed.

vi. And that this dynamic represents what may be called a healing instinct, or instinct to wholeness, that certainly operates on every level of human experience, and possibly at earlier and deeper stages of evolution.

The end of evolution

It has been suggested that evolution reached its climax and fulfilment in the life of Jesus (Freeman 2000 p.256), a question also considered by Fraser Watts (Chapter Four of this book). This was the point at which it was timely for human life to be united with its divine origin. If this is so, and if, as St Paul implies, the full maturity of the human personality lies in letting the same mind be in us as in Christ Jesus (Philippians 2.5), then the Christian vocation is the point at which Theology, Evolution and the Mind converge. And if, as Austin Farrer proposes, "there may be, so to speak, a Christ factor where there is no Christ", by extension this point of convergence is in principle common to all humanity (Farrer 1994 p.107).

My 40-plus years of clinical and pastoral experience has convinced me that this convergence is exemplified in the search for wholeness in individual lives. There is an instinct for wholeness that finds expression in experiences of human healing, and that is reflected in the physiological, psychological and spiritual well being of individuals (Swayne 2005).

The psychosomatic whole

For many years I have been dealing mostly with chronic and complex heath problems, and it has been increasingly apparent that whatever the presenting problem and its physical and psychological components, there is surprisingly often an element of the story that has something to do with the person's spiritual life. Either with experience of religion – often, sadly, bad experience – or with some unsatisfied sense of God-consciousness; or with some lack of meaning or value that has to do with more than worldly needs; or with some sense of unfulfilled potential in the person's life that I would call vocation, and which I find to be inseparable, at least in some small degree, from that sense of God-consciousness.

I have come to believe that the striving for integrity and wholeness on this level is as much an inherent instinct as the activity of the body's self-regulation and repair mechanisms in the face of physical damage and disease (Swayne 2005).

These mechanisms and this healing instinct have common characteristics at whatever level of our being – body, mind or spirit – they operate; such as the new growth of tissue repair, psychological insight or spiritual awareness (Swayne 2005).

The common characteristics of these processes are represented to some extent in the Table below, but here I want to point out not only their necessity to our present physical, psychological and spiritual well-being as, perhaps, fully evolved human beings, but also their role throughout evolution.

	BODY	MIND	SPIRIT
Awareness	Immune response	Dis-ease, Breakdown	No meaning in life
Conditions for healing	Wound care, Nutrition	Empathy, Understanding	Soul friend, Discernment
Mobilising resources	Inflammation, Self-regulation	Truthfulness, Willingness	God-consciousness
New growth	Tissue repair	Insight, Self-worth	Spiritual commitment
Reconciliation	Adjustment, Compensation	Rebuilding relationships	Forgiveness, Hope

Healing as an evolutionary imperative

Regardless of other attributes that equipped it to compete successfully and establish itself, no organism would have survived without the capacity to resist and to recover from the hostile influence of its environment and its competitors, and from disorder within itself. Preserving health, whether by protective and prophylactic means or by healing processes is an evolutionary imperative. This imperative, present inevitably from the first emergence of life, and perhaps in some sense before it, is also now to be seen in this deeper response to our evolutionary goal of becoming a completely realised human being.

My attempt to reconcile this insight with what I have understood of subtle evolutionary processes from the work of John Polkinghorne (Polkinghorne 1998, 2005) and Simon Conway Morris (Conway Morris 2005), and, ironically perhaps, Richard Dawkins (Dawkins 2006a), has lead me to propose three principles or dynamics on which the momentum of evolution depends. These are functionality, integrity and relationship.

Functionality gives the organism its competitive edge, its ability to perform and reproduce successfully, its "fitness for purpose". Integrity (integratedness or wholeness), a composite ecological value – its place in the bigger picture. Relationship determines the working out and fulfilment of its role, or at least in a metaphorical sense, its "vocation". The ecological value of an organism resides in its relationship with others; the manner in which it competes or cooperates with them. Those relationships will in turn affect its own evolutionary progress and the progress of, so to speak, its "neighbours".

These three dynamics can be held not only to have determined the progress and outcome of past stages of evolution, including gene formation, and the present state of all organisms, but also to determine the fulfilment of our potential, and indeed our survival, as human beings now. And all three both determine and depend upon the heath maintenance and healing processes of which I have spoken. There is, then, a kind of circular relationship between the dynamics of evolution and the dynamics of healing and wholeness. But it is not a closed circle. It is an entwined spiral. Another Caduceus.

This paper is an attempt to formalise for the first time ideas that have been born in me by the convergence of my experiences of human healing and my recent acquaintance with the conversation between science and religion. If they interest you sufficiently and you can offer any encouragement, guidance, or indeed correction, I shall be grateful.[2]

Postscript

In discussing this paper with a friend and mentor, I was challenged to consider whether evolution in itself has made things better[3]. Is human behaviour on the whole better than animal behaviour? Is the behaviour of modern man on the whole any better than that of primitive man? Perhaps the answer is that evolution is not about making things better; except in the sense that, as Richard Dawkins would have it, evolution has produced progressively more efficient genetic replicators (Dawkins 2006b p.130). It is, rather, about developing complexity and diversity. And it is certainly about the emergence and development of consciousness and knowledge – of human nature and of the whole creation. But these things have not necessarily in any convincing fashion made things on balance better. What they have done is to create an expanding repertoire of opportunities and challenges. Out of these have come many good things and many bad

---

[2] E-mail to "JEREMY SWAYNE" <jem.swayne@btinternet.com>
[3] Note the several other discussions of the concept of evolutionary progress in this volume: notably in Chapters Four and Twelve. Ed.

things. And the supreme opportunity and challenge, a Christian might say, is to bring in the Kingdom of God.

Evolution, Dawkins says, raises a question: "The world is full of things that exist . . . ! No disputing that, but is it going to get us anywhere?" (Dawkins 2006b p.127). It might be correct to agree with him that regarding the biological process itself, "Evolution has no goal" (*Ibid.* p.50), but it could be true to say that on the other hand evolution "serves a purpose". We could say that whatever initiated the birth of the universe, establishing conditions conducive to evolution, may have had a long-term goal. It is impossible either to confirm or refute that speculation, because we do not know what initiated the "big bang"; but the fact that evolution has produced that repertoire of opportunities and challenges, which are certainly consistent with the possibility of some purposeful aspiration or meaning in life, perhaps allows us to infer it.

This speculation bears upon the question referred to earlier concerning the birth of Christ as the end of evolution. We might argue that although the incarnation and the new "testament" initiated a new beginning in the human spiritual journey, the evolutionary journey – the repertoire of opportunities and challenges necessary to this new departure on the human journey towards God, made possible by the emergence of human personality, consciousness, self-awareness and God-consciousness – was complete. No new resources were needed, only the best and proper use of those already available; new insights and attitudes, a *metanoia*[4] certainly, and a new willingness to aspire to and attain the union with God that Jesus taught and proclaimed, but no new evolutionary tools.

References

Brooke, J.H. 2007. Arthur Peacock: An appreciation. *Reviews in Science and Religion*, 49, 7-12.
Conway Morris, S. 2005 *Life's Solution*. Cambridge: University Press.
Dawkins, R. 2006a *The Selfish Gene* (edn 3) Oxford: University Press [1st edn 1976].
—. 2006b *The Blind Watchmaker,* London: Penguin [First published London: Longman 1986].
Freeman, L. 2000. *Jesus the Teacher Within.* London & New York: Continuum.
Farrer, A. 1994. *Saving Belief,* London: Mowbray; Harrisburg: Morehouse.

---

[4] metanioia = change of mind, repentance. Ed

Polkinghorne, J. 1998. *Belief in God in an Age of Science.* New Haven & London: Yale University Press.
—. 2005. *Exploring Reality.* London: SPCK.
Swayne, J. 2005. Homeopathy, Wholeness and Healing. *Homeopathy* 94(3), 37-43.

CHAPTER SEVENTEEN

THE MIND IN THEOLOGICAL AND SCIENTIFIC PERSPECTIVE

SJOERD L. BONTING

Revd Prof Sjoerd Bonting studied biochemistry in Amsterdam (PhD 1952). He worked in the US for the next 13 years, returned home to the Chair of Biochemistry at Radboud University, Nijmegen (1965-85), and continued as consultant to NASA's preparation of the International Space Station till 1993. While in the US he also studied theology and was ordained into the Episcopal Church. Back in the Netherlands he founded and served four Anglican chaplaincies, and still writes extensively on the theological challenges of modern science. His books include "Chaos Theology" (2002) and "Creation and Double Chaos" (2005).

Here he offers a concluding contribution to this book. Though too long to be included among the contributed papers in the 2007 conference, it is in fact written with characteristic succinctness. It is the paper's coverage that is the cause of its length, for Prof Bonting sweeps magisterially through at least two themes of previous speakers – the evolution of religion and the philosophy of mind – as well as presenting two entirely new topics: biblical views of mind and soul, and the neuroscientific account of the brain. Though we are obliged to refer readers to the Web for visual illustrations of the latter section, and not all aspects of Prof Bonting's overview of the subjects treated earlier by Profs Mithen and Trigg tally with their positions, his tour-de-force coverage can hardly fail to provide each reader with a challenging finale to our symposium.

Considering the mind in theological and scientific perspective is fraught with the difficulty that we must look with our minds at "the mind", which makes it hard to avoid subjectivity. Another problem is that the biblical data on the mind are rather scarce, though there is a clear implication of body-mind unity, while the theological treatment – under the influence of Greek philosophy – became centred on the soul and its moral aspects. Philosophy and psychology, in the absence of scientific data, were mired in the dualist-monist controversy.

Neuroscience, with the aid of brain scanning techniques, has provided extensive understanding of the biological substrate of the mind, confirming the biblical body-mind unity. An evolutionary development of brain structure and function can be traced. All human mental capabilities and traits are found to be present, at least to some extent, in non-human primates. Thus, it is impossible to indicate a moment of change that can be called "hominization". Religion also shows an evolutionary development that can be interpreted as the interaction between the human mind and divine revelation. Death and resurrection are considered here in the light of these findings.

As a tentative definition of the mind I propose:

> The mind is the complex of faculties involved in the processes of perceiving, remembering, considering, evaluating, deciding, emotion, and religious experience (*Brittanica* 1989 p.151). Consciousness can be regarded as the living mind at work.

Subjects considered in this chapter are: biblical view (1), theology, philosophy and psychology (2), neuroscience (3), evolution of the mind (4), evolution of religion (5), death and resurrection (6), and, in conclusion, a discussion of the findings (7).

1. Mind in the biblical view

Old Testament (OT)

The Hebrew understanding of mind and thinking differs from ours (Dentan 1962 pp.383-4; Robinson 1913 pp.4-67). First, there is no abstract thinking for its own sake, nor any psychological analysis. Secondly, no specific anatomical location is assigned to the mind; thinking, planning, willing, and feeling are thought to be functions of the entire personality, the body-mind unity.

Where the Hebrew word *leb* (literally: heart) is translated as "mind" in the New Revised Standard Version (which is used throughout this essay), it can refer to the seat of recollection (Isa.65:17; Jer.3:16), to will and purpose (Isa.26:3; Jer.19:5), memory (Deut.4:9), thinking related to past or future action (1Sam.2:35; Ezek.11:5), technical skill (Deut.4:9); and love (2Sam.14:1). In other cases the literal translation "heart" is used, where "mind" would also apply, to indicate personality, inner life, character

(1Sam.16:7; Gen.20:5); emotional states such as joy or sorrow (1Sam. 1:8), anxiety (1Sam.4:13), and fear (Gen.42:28); intellectual activities such as attention (Ex.7:23), understanding (1Kgs.3:9); and will or purpose (1Sam.2:35).[2]

In a few cases the words *kelayoth* (kidneys), *meim* (bowel), and *kabed* (liver) are used in a similar sense. The brain is never mentioned; it was considered merely as "the marrow of the skull". There is no Hebrew word for body, but *basar* (flesh), its nearest equivalent, is sometimes used in a psychological sense: indicating fear (Ps.119:120, Job 4:15) or enjoyment (Eccles.2:3), as influenced by the mental state (Prov.14:30: *a tranquil mind gives life to the flesh*). All this indicates that the Hebrews considered the mind as an aspect of the body without seeing it located in any particular organ.

Then there is the question: What makes the mind come alive? For this "life-principle" the OT uses the three virtually synonymous Hebrew words *nephesh, neshamah* and *ruach*. The first of these, *nephesh* is translated as "life" (1Kgs.19:10), as the "soul" leaving at death (Gen.35:18) or as "breath" returning to God who gave it (Eccl. 12:7). It is interpreted as the personal "yourselves" (Ezek.4:14; Lev.11:43) and as the emotions "anguish" (Gen.42:21) and "desire" (Deut.21:14). *Neshamah* is translated as "breath" in the sense of life (1Kgs.17:17; Job 27:3). *Ruach*, which has the three-fold meaning of wind, breath and spirit, is used as "wind" in Ex.10:13, as "spirit" in Gen.45:27, Judg.15:19, 1Sam. 30:12, 1Kgs.10:5, and Hag.1:14, and as Yahweh's "breath" in Ex.15:8 and Isa.30:28. It is also used for emotions as anger (Judg.8:3; Gen.27:45) and grief (Gen.26:35). For God's life-giving breath, *neshamah* is used in Gen.2:7, but *ruach* in Ezek.37:9-10. In this sense, *ruach* is used as inspiring prophecy (1Sam.10:3,6) or extraordinary strength (Judg.14:6), and its loss as causing insanity (1Sam.16:14). In Psalms and Proverbs *ruach* is employed as a synonym of *nephesh* for the "inner life". In Isa.26:9 *ruach* is used in parallel with *nephesh*: "My soul [nephesh] yearns for you in the night, my spirit [ruach] within me earnestly seeks you."

From this survey of OT texts it appears that the word *leb* on the whole represents the human mind in its various activities, while the words *nephesh, neshamah* and *ruach* represent primarily the life-giving principle, God's breath that makes the human person come alive and that is withdrawn at death. There is, however, some overlap between *leb* and the other three words, particularly when it comes to emotions. It also appears that no clear distinction was made between mind and soul.

New Testament (NT)

The NT basically follows the OT line in seeing human personality as a unity of mind and body, particularly in the synoptic gospels and the Pauline writings (Dentan 1962 pp.383-4; Robinson 1913 pp.68-150). The latter provide the most extensive treatment of the subject. For Paul the mind is the thinking, reasoning, reflecting and purposing aspect of the human self, while the body (*soma*) is the same self as the object of these activities. The mind enables us to comprehend God's revelation and to act upon it. Being human, the mind is capable of being corrupted (Rom.1:28; 1Tim.6:5), but also of recovery (Rom.12:2). For Paul "mind" is in some sense the whole human being and can often be taken as equivalent to "character".

Paul, writing in Greek, uses the word *nous* (mind) as an equivalent of the OT *leb* (Rom.11:34, 1Cor.2:16). It is the intellectual faculty of natural man, which can be morally good as well as bad (1Cor.14:14, Phil.4:7). Paul also frequently uses *kardia* (heart) in the sense of personality, character, inner life (e.g., 1Cor.14:25), and as the seat of intellectual activities (Rom.1:21), volition (Rom.2:5) and emotional states of consciousness (Rom.9:2).

For the term *nephesh* Paul uses the word *psyche*, and for *ruach* the word *pneuma*. Of these, *psyche* is translated as "life" without psychological content (Phil.2:30), "anguish" (Rom.2:9), "heart" (Eph.6:6) and "soul" (1Thess.5:23). In the last three texts *psyche* stands for the emotional side of consciousness. *Pneuma* connotates mostly "supernatural influences", rarely "principle of life". Frequently, it is used for "human spirit", but in Rom.8:16 Paul uses it both for God's Spirit and the human spirit ("that very Spirit bearing witness with our spirit").

In the synoptic gospels there is – compared to Paul – a change of emphasis rather than of content. Jesus teaches the need of a spiritual life (Lk.10:38-42) rather than an ascetic life (Mt.11:19), the weightier matters of the law are justice and mercy (Mt.23:23), the forgiving of sins (Lk.19:1-10), and the life beyond death. John has Christ bring the light of life to a world of darkness (Jn.8:12, 12:46). A distinction is made between sinfulness as a character attitude (Jn.8:34) and a single act of sin that can be forgiven (1Jn.1:9). This requires moral volition on our part (1Jn.3:3).

"Soul" is a confusing concept (Porteus 1962 pp.428-9). In the OT it is essentially the life principle, which at death is thought to depart (Gen.35:18). However, it is also used for the self as the subject of appetite and emotion, where also "mind" could be read. In the NT it can mean "life" that can be cared for (Mt.6:25), saved or lost (Mk.8:35), laid down (Jn.10:11), but it can also be the subject of emotion (Mk.14:34; Lk.1:46).

Thus "soul" varies between "mind" and life-giving "spirit" (*leb* vs. *ruach*; *psyche* vs. *pneuma*). For this reason, and also because it is frequently, but wrongly from a Christian point of view, associated with immortality, I shall avoid the further use of the term "soul" and include religious experience among the activities of the mind.

Summarizing, we may say that in the biblical view "mind" (*leb*; *nous*, *psyche*) is the thinking, reasoning, recollecting, planning, willing, and feeling aspect of the human self, enabling knowledge of God, capable of corruption as well as recovery. Mind, and thus the whole person, comes to life upon receiving God's life-giving "spirit" (*ruach*, *nephesh*; *pneuma*), which is withdrawn at death.

2. Mind in theology, philosophy and psychology

Theology

In the patristic and scholastic periods discussion was almost entirely centered on the "soul" and its involvement in sin and salvation. The thinking was strongly influenced by Greek philosophy, Stoicism, Platonism and Aristotelianism, in succession.

Tertullian (c. 200), under stoic influence, saw the "soul" as an entity with many functional activities, the rational and volitional (*nous*, mind) being the highest; it is centred in the heart, but the body is simply its instrument. **Clement** (c. 200), under Platonist influence, believed in a tripartite "soul", consisting of an intellectual part (*nous*, considered divine and immortal), a part for passion, and one for desire. **Origen** (c. 225), also a Platonist, assumed a unity of body, soul ("mind"), and spirit, close to NT thinking. **Gregory of Nyssa** (c. 370) has the Aristotelian form of a soul with vegetative, animal, and intellectual parts, independent of the body. **Augustine** (354-430) adopted Origen's idea of the unity of body, soul ("mind"), and spirit. He claimed that the soul is directly created by God, but is passed on from parent to child. This idea led him to affirm the doctrine of original sin.

Thomas Aquinas (1225-74) held that the soul is created by God and placed in the body, but is not transmitted to the descendants. The soul becomes capable of acquiring merit only upon the infusion of divine grace. Aquinas accepted free will, but "free will cannot be converted to God, unless God converts it to himself." Over against Aquinas' emphasis on reason, **Duns Scotus** (1264-1308) emphasized the will, both divine and human. The human will is to keep order in the rebellious constitution of humans but is corrupted by the Fall. He saw humans as a unity of body and soul in a unique individual, with the soul capable of knowing the spiritual intuitively.

The Reformers contributed little to the understanding of the human mind. **Martin Luther** (1483-1546) preached *sola fides*, justification by faith alone. For him faith was full trust in God, rather than intellectual assent. He maintained the doctrines of original sin and predestination. **Philip Melanchthon** (1497-1560), the main theologian of the Lutheran movement, nuanced *sola fides* by assuming the need for cooperation between God's Spirit and human will in conversion. **John Calvin** (1509-64) maintained predestination, original sin, and loss of the free will after the Fall. Emphasis on God's omnipotence and neglect of God's mercy, led Calvin to the doctrine of "double predestination", i.e. before creation God already predestined some of his creatures to salvation and others to damnation.

Philosophy

Philosophy took over the problem of the mind from theology during Renaissance and Enlightenment and the arrival of modern science (Wikipedia 2008). This period was marked by "rationalism", a belief in the all-sufficiency of human reason, in opposition to all supernatural belief and dogmas.

The *mind-body problem* was a major issue in the struggle between dualism and monism. *Dualism* is the idea that mind and body belong to different and separate categories (**René Descartes**, 1596-1650). *Monism* is the position that mind and body do not exist as distinct types of entity (**Baruch Spinoza**, 1632-77). Since dualism is incompatible with both the biblical view and the findings of neuroscience, I shall confine myself to monism. Three types of monism can be distinguished: 1. *physicalism* which maintains that the mind consists of physical entities and is not something different from the body; 2. *idealism* which claims that the mind is all that exists and that the external world is either mental itself, or an illusion created by the mind (**George Berkeley**, 1685-1753); 3. *neutral monism* which holds that there is some neutral substance, of which both matter and mind are properties (**Spinoza**). Idealism and neutral monism seem to be incompatible with both the biblical view and current neuroscientific insights, so will not be considered further here.

This leaves physicalism, which has been the dominant idea in the last century (Horgan 1994 pp.471-479). *Reductive physicalism* asserts that all mental states and properties will eventually be explained in terms of physiological processes and states, which is incompatible with biblical thought. *Non-reductive physicalism* recognizes that the brain is the substrate of the mind, but acknowledges that mental processes are not identical with the neuronal processes observed in neuroscience. Between them there is a "property dualism", as some call it.

Other types of physicalism are: behaviorism, identity theory, and functionalism. *Behaviorism* focused on behaviour because of the difficulty of studying introspection, but is out of favour now. *Identity theory* (**J.J.C. Smart**, 1956) operates on the premise that mental states are identical to the firing of certain neurons in certain brain regions (Levina *et al.* 2007). However, a mental state like pain can occur in many different species, and it seems unlikely that in all cases the same brain state would operate. This led to *functionalism* (**Hilary Putnam**, 1967), which characterizes mental states by their causal relations with other mental states and with sensory inputs and behavioral outputs (Nicolelis & Ribiero 2006). For example, a kidney is characterized scientifically by its functional role in excreting waste products and retaining desired substances; for a functionalist it doesn't matter how the kidney is made up; its role and its relation to other organs define it as a kidney. Linguistic philosophers, however, claim that trying to fit mental and biological states together is a "category error". **Ludwig Wittgenstein** (1889-1951) said:

> ... we use the word mind without any difficulty until we ask ourselves "What is the mind?" We then imagine that this question has to be answered by identifying some "thing" that is the mind.

Psychology

How then did psychology, the science of the mind (*psyche*), deal with the problem? First **introspection** was used, but this was found to be too subjective to arrive at predictive generalizations and thus to provide a theory of the mind. Then the attention turned to **behaviour** as the expression of mental states, without making any assumptions about the mind or even its existence. When behaviorism was found not to provide further insight in the mind, it was succeeded by **cognitive psychology**, in which the mind is seen as an information-processing system, which permits the use of computational models to study the functional organization of the mind. Again, this did not produce much further insight into the mind itself before use was made of the findings of neuroscience.

It seems clear that theology, philosophy and psychology have not provided a deeper understanding of the mind and its activities, basically because of the lack of data and in the case of theology its concentration on the soul and its moral aspects. This changed with the advent of noninvasive scanning techniques in the 1980s. These have provided high resolution structural and functional information on the living brain, which has played a leading part in the development of a new branch of science, called neuroscience.

3. Mind and neuroscience

In the next three paragraphs I provide, for the benefit of readers unfamiliar with this, a brief description of the nervous system and the modern techniques employed in studying it.[1]

Nervous system

The brain is the control centre of the nervous system, which is made up of nerves traversing the body to the brain (afferent) and from it (efferent). It consists of over 100 billion nerve cells (neurons), each linked to up to 10,000 other neurons. The cerebellum ("small brain") is the site of integration of sensory perception and motor control. Neural pathways connect it with the cerebral motor cortex, which sends commands to the muscles. The spinal cord, an extension of the brain down into the vertebral column, transmits nerve signals from the periphery to the brain (touch, position sense or "proprioception", pain, temperature) and vice versa. The mind is relieved by the fact that many functions are performed autonomously (unconsciously), such as the regulations of blood pressure, fluid balance, body temperature, and breathing. The high activity of the brain is indicated by the fact that it utilizes 20% of the body's total energy consumption, although it represents only 2% of the body weight.

A neuron has a cell body with short projections ("dendrites") in one direction and a long one ("axon") in the opposite direction. Each axon connects to a dendrite of the next neuron in the chain at a point of close contact called a "synapse". A nerve signal is conducted along the axon as

---

[1] Illustrations of the brain and neural structures can be found at many websites. *http://www.morris.umn.edu/~ratliffj/images/brain_slides* provides informatively captioned starting diagrams – slides 1,5,9 are suggested; *http://neuroanatomy.ca* has a wide range of diagrams (e.g. Limbic System, CH-7) plus real cross-sections under Web Atlas/Cross Sections/Coronal – explore 6A-8B, clicking "Key" in each case. Individual structures can be located on high quality MRI scans at: *http://www.med.harvard.edu/AANLIB/cases/caseNA/pb9.htm*: tick all 3 "show" boxes, then select particular structures in turn from the tabulated list. For greatest detail of all, though displayed in varying ways, see:
http://en.wikipedia.org/wiki/List_of_regions_in_the_human_brain. Ed.

an electric spike (or "action potential"). When the spike arrives at the synapse, it causes the release of a small burst of a so-called "neurotransmitter" chemical which diffuses, typically in less than a millisecond, across the very narrow gap between the two cells and triggers a spike in the next neuron (or, at the end of a neuron chain, in a muscle fibre or gland cell). Because only the incoming nerve terminal contains transmitter, and only the receiving cell membrane can respond to it, synaptic transmission can only go in one direction; thus the synapses impose the direction of information flow in the nervous system. After transmission the neurotransmitter is removed from the synaptic cleft by reuptake or by degradation. Examples of neurotransmitters and their actions are:

- Acetylcholine – activation of the voluntary muscles
- Noradrenaline (or norepinephrine) – wakefulness or arousal
- Dopamine – motivation, desires
- Serotonin – memory, emotions, wakefulness, sleep, temperature regulation
- Gamma aminobutyric acid (GABA) – inhibition of motor neurons
- Glycine – spinal reflexes and motor behaviour.

Because of their diverse and exquisite chemical sensitivities, it is at the synapses that the majority of drugs affecting the nervous system operate.

Recently, it has been shown that the synapse is a so-called "non-linear system", which makes synaptic transmission subject to "critical avalanches" or "chaos events" (Levina *et al.* 2007).

Techniques

Detailed knowledge of the neural circuits that operate in the various functions of the brain has become available through the application of the following techniques:

- Electroencephalography (EEG): available since the 1920s, this provides high temporal resolution (milliseconds), but poor spatial resolution (shows activity of many thousands of neurons firing synchronously).
- Magnetic resonance imaging (MRI): 2- or 3-dimensional images with a high degree of anatomical detail, produced by radio wave emission from atomic nuclei responding to changes in a very strong magnetic field.
- Functional MRI (fMRI): measures neural *activity* in specific regions of the brain, via changes in blood flow due to the metabolism associated with neuronal activity, e.g., in synaptic reuptake of neurotransmitters.

- Positron emission tomography (PET): monitors glucose metabolism and neurotransmitter activity in specific brain regions. Combination with a computerised tomography (CT) scan provides precise location. It requires injection of a positron-emitting substance, usually fluorine-18 labelled deoxyglucose.

The last three techniques, which became available in the past twenty years, provide spatial resolutions of 1 mm and time resolutions of minutes. All the foregoing are classified as "imaging techniques", but important information has also been obtained by:

- Transcranial Magnetic Stimulation (TMS): a technique to stimulate neurons from outside the brain.

After this brief account of the nervous system and the techniques employed in studying it, I now describe some of the findings obtained in recent years.

Neural network interaction

Mental activities are carried out by ensembles of many neurons, linking distinct (and distant) regions of the brain. They synchronize their activities, with tasks requiring much effort leading to low-frequency responses of about 28 Hz (1 Hz = 1 cycle/sec), and effortless tasks to higher frequency responses, about 45 Hz (Knight 2007). Episodic psychiatric disorders without visible anatomical damage, such as depression, might be due to loss of synchronicity in a neural network.

Personality

Different patterns of brain activities are found in different personalities. Optimists have a high activity in the rostral anterior cingulate cortex and the amygdala, parts of the limbic system that processes emotion and memory (Schacter & Addis 2007). Progressives have a higher activity in the anterior cingulate cortex than conservatives (Amodio *et al.* 2007). Depressed persons show reduced coupling between the prefrontal cortical area and the amygdala. Extroverts show a high activity in the orbito-frontal cortex. Neurotics have a high activity in the insular cortex. In drug addicts the synapses in the mesolimbic region are flooded with dopamine, by increased release (amphetamines), by blocking of reuptake (cocaine), or by blocking of GABA release (heroin, morphine; GABA inhibits the dopamine effect) (Kauer & Malenka 2007). Addiction is due, in part, to powerful memories of the drug experience but also to an increased number of dopamine receptors. Relapse is facilitated by a decreased activity of the prefrontal cortex, which would normally function to control compulsive behaviour.

Mirror neurons

These are paired neurons, a "Self" neuron and an "Other" neuron. The latter permits us to experience what others see, feel, experience and think, to imitate others and to have "empathy" (Rizzolati & Criaghero 2004). In songbirds, mirror neurons fire when another bird sings a song similar to their own (Miller 2008). The "Self" neuron is repressed, when the "Other" neuron is activated, such as when a monkey sees another monkey perform the same action. In children mirror neurons are probably involved in learning and language acquisition. In autistic persons the mirror neurons for action, feeling and thinking are dysfunctional. In schizophrenia distinction between "Self" and "Other" may be impaired, leading to delusions and hallucinations, e.g., ascribing "voices" to "others".

Plasticity

Neurons grow or disappear depending on age, circumstances and training. This "plasticity" is most pronounced in early life, but recent evidence shows that new neurons grow from stem cells in the adult brain. Such "neurogenesis" occurs in specific regions such as the cortex, hippocampal dentate gyrus and olfactory bulb (Scharfman & Hen 2007). Another type of plasticity is the partial takeover of activity which has been blocked in one region (e.g., after obstruction of a cerebral vessel by a blood clot) by another region. This may explain why rehabilitation and psychotherapy have favourable effects on cerebral structure and function even at advanced age. In Parkinson's disease, plasticity is lost through the death of dopamine-using ("dopaminergic") neurons. Both dopaminergic drugs and deep brain stimulation of the subthalamic nucleus can at least temporarily halt this process, though they act by different circuits (Frank *et al.* 2007). Chronic stress causes loss of nerve connections in the hippocampus and an increase in the number of dendrites in the amygdala.

Memory

Three types of memory are recognized: (1) Sensory memory is established within 0·2-0·5 sec after perception of a maximum of 12 digits, and is lost in about the same time; (2) Short-term memory is established by transfer of sensory memory, but is normally considered to be limited to

5 digits; it lasts for minutes and involves transient neuronal patterns in the frontal and parietal cortex; (3) Long-term memory results from the conversion of short-term memory in the hippocampus (but is not stored there) by repetition or training and lasts from 3 months to lifetime; it involves plasticity of neuronal connections and is consolidated during sleep (Nitz & Cowen 2008). Suppression of emotional memories appears to occur by a two-phase neural mechanism (Dupue *et al.* 2007). Damage to the hippocampus may cause memory deficit accompanied by difficulty in imagining the future (Miller 2007). The basal ganglia, in cooperation with the prefrontal cortex, appear to act as a filter to prevent remembering irrelevant information (McNab & Klingberg 2007).

Self awareness

This includes awareness of body, of memories and of place in society. The medial prefrontal cortex, in cooperation with the precuneus, seems to bind together all memories and perceptions relating to oneself, a role similar to that of the hippocampus in creating memories (Zimmer 2005). In Alzheimer's disease the first damage occurs in precuneus and hippocampus, both involved in the formation of memory and of images of past and future. In the vegetative state due to brain damage there is no awareness (only reflex movements, not on command). In a woman in vegetative state, studied with fMRI, spoken sentences triggered activity in the superior and middle temporal gyri (as occurs in understanding speech); asking her to imagine walking through her house elicited activity in the networks involved in spatial navigation, the premotor, parietal and parahippocampal cortices (Laureys 2007). These observations were made while she had no consciousness whatsoever; much later, a year after her accident, she had become minimally conscious with a fair chance of recovery.

Free will

In a famous experiment, Benjamin Libet had people make a voluntary hand movement while watching a specially rapid clock (full rotation in 2·5 sec) and then indicate at what clock time they were aware of their intention to make the movement. They also had electrodes on their heads recording activation of their motor cortex (the "readiness potential"). The readiness potential surprisingly preceded awareness by 500 msec; movement followed awareness by 200 msec (Libet 1985). Some conclude from this experiment that there is no free will in "willing" a movement, though there could be in stopping the movement.[2] A better explanation is that there is a half-second delay between our making a decision and becoming conscious of having done so.

HPA axis

The mind interacts with the body, not only by peripheral nerves, but also by means of the so-called hypothalamic-pituitary-adrenal axis. Neuronal impulses travel from the cerebral cortex to the hypothalamus, which then secretes certain compounds to the pituitary gland, which in turn releases certain hormones that stimulate the adrenals to secrete other hormones. In this way the mind can influence the action of various physiological functions. For instance, various types of stress affect the immune system, raising vulnerability to infection and cancer (Segerstrom & Miller 2004). Conversely, the condition of the body can influence the mind, e.g., inflammations can lead to depression through stimulation of the immune system (Dantzer *et al.* 2008).

4. Mind in evolution

Early multicellular organisms had a single nerve cord that extended through the body of the animal. In arthropods, such as insects, a primitive brain was formed from the nerve cord, comprising a three-part cerebrum and optical lobes. In vertebrates the nerve cord changed into the spinal cord, and the brain then comprised the cerebrum, cerebellum (each with a left and right hemisphere) and spinal cord (Butler 2000). In mammals the neocortex, the top six-fold layer of the cerebrum, appears. In primates the neocortex is characterized by its foldings and large prefrontal cortex. The latter is the seat of the higher functions, such as sensory perception, motor commands, spatial sense, and, in humans: conscious thought, and language. Apart from a threefold difference in volume (adult human 1350 cc, chimpanzee 385 cc), the microscopic structure of human and primate brain differs only in some minor details for which no functional significance is known (Balter 2007).

Neither are there great differences between the human and chimpanzee genomes. They differ in only 1·2% of their 2·4 billion base pairs. The differences represent genes involved in smell, hearing and protein breakdown, but these are hardly relevant for mental capabilities (Gunter 2005). More important is the degree to which genes are co-

---

[2] Cf., Choong, Chapter Fifteen of this volume. Ed.

expressed with other genes. When such co-expressed genes are grouped in networks, it is seen that many connections noted in humans are missing in chimps, as many as 17·4% of them in the cortex (Cohen 2007b).

How can we explain the superior mental capabilities of humans as compared to non-human primates? Actually, many of these differences are not as great as we may think. Many traits that seem to be peculiar to humans are found in chimpanzees, other monkeys and even lower animals (Cohen 2007a). Chimpanzees, our nearest relatives, are social creatures capable of empathy, altruism, self-awareness, cooperation in problem solving and learning through example and experience. They make simple tools such as stick spears to hunt smaller primates for meat, use stones to crack nuts and sticks to extract termites from their nests, fold leaves to collect water from tree hollows, scoop algae from stream surfaces, and pound juicy palm fibre to pulp for food. Young chimps watch nut-cracking by adult chimps, learning by example. Learning begins at age 1 yr, and at age 3 to 5 they can master three-object tasks. At age 3 they learn to stack blocks; humans do so at age 1. Two-month old chimps make eye contact with the mother, after 1 yr (the same age as for humans) they are able to maintain the gaze as the mother moves around. Young chimps outperform college students in some memory tasks, such as selecting the randomly and briefly displayed numbers 0 through 9, in ascending order on a touch-screen (students correctly selected only 4). Rhesus monkeys, after training, did as well as college students in adding up to 20 dots. Both these findings suggest an evolutionary continuity of basic mathematical skills in humans and other primates (Ledford 2007).

Chimps cooperate in obtaining food from the other side of a fence. Bonobos, a dwarf chimpanzee species, do even better in cooperative tasks and share food more readily. Chimpanzees show empathy, e.g., in consoling the loser in a fight, mediating in a conflict, grooming an animal with cerebral palsy, and in a mother carrying her dead child for days on her back.

Does language capability distinguish humans from the animals? The *FOXP2* gene that is needed for all vocalizations, whether shrieks, songs or sentences, is present in all vertebrates (Whitfield 2008). It produces a protein that binds to the DNA of some 100 genes, switching some on and others off. The *FOXP2* gene is thought thereby to coordinate the 100 or so muscles of face and mouth involved in producing sounds or speech. It appears that human language capability has arisen gradually, rather than suddenly, from precursors in earlier primates (Ghazanfar 2008).

The statement in Gen.1 that humans are God's image bearers, which is not said of any animal, would seem to be in line with the fact that humans and chimpanzees show an opposite time-course of mental development. At birth they have about the same brain size and perform about equally in various tests, but after about 18 months the human baby starts to leave the chimpanzee far behind in brain size and mental performance.

Various factors may have contributed to the evolution of the human brain:

- Bipedalism: Originating in *Australopithecus*, this freed the hands for grasping food and other objects, while relieving the jaw from such tasks. This permitted the development of throat, tongue and mouth, which eventually enabled language. The use of the hands for food gathering and defence required coordination with the eyes, leading to increased cerebral volume and complexity. Narrowing of the pelvis caused earlier birth, necessitating the carrying of babies in the mother's arms, thus leading to stronger bonding;
- Diet: The invention of fire making, possibly 200,000 yrs ago by *Homo erectus* or else 165,000 yrs. ago by *Homo sapiens*, introduced eating of cooked food (Minkel 2008). The energy saved on digestion contributed to the expansion of the human brain.
- Social life: Living in families began among the particular species of *Australopithecus* called *Australopithecus afarensis*, as witnessed by the "First Family" of 13 individuals, whose fossils were found in Ethiopia. The challenge of living in a social group stimulated the evolution of primate intelligence, as shown by the fact that baboons, who live in families, clearly distinguish between "me" and "not-me" when hearing a playback of baboon calls (Cheney & Seyfarth 2007), and that neocortical volume (relative to total brain volume) increases with primate group size (Dunbar & Schultz 2007).
- Cultural development: Chimps, macaques and tamarins, like human infants, have the ability to distinguish goal-directed from random action (Pennisi, 2007). Children at age 2·5 yrs. have the same cognitive skills as chimpanzees for dealing with the physical world, but have more sophisticated cognitive skills for dealing with the social world (Hermann *et al.* 2007). Cooperative behaviour in non-human primates is mainly limited to kin and reciprocating partners, and is virtually never extended to unfamiliar individuals (Silk *et al.* 2005). When a monkey dies, cortisol levels in closely-related females rise, indicating stress, but no such response is detectable in other females of the group (Silk 2007). A specific type of tool use, taught to a single chimpanzee, spread to 30 of 32 group mates, but not to an unrelated group (Whiten *et al.* 2005). The extinction of the Neanderthals, who were sophisticated toolmakers and good hunters, has been

ascribed to a lack of the working memory that permitted *Homo sapiens* to solve novel problems on the strength of different "wiring" of their prefrontal and basal ganglia (Wynn & Coolidge 2008). Another possibility is that Neanderthals lacked the analytic language that *Homo sapiens* had mastered (Mithen 2005, and Chapter One of this book).

5. Mind and the evolution of religion

When did religion originate and how did it develop? The first question is difficult to answer for a period without written documents. Burial with signs of ritual is generally taken to indicate a beginning of belief in life after death. The earliest case widely interpreted in this way is a Neanderthal burial site, dating from about 100,000 years ago, of a child with a small stone on the skull and one on the heart region, which were of a type of rock found only 100 km away from the place of burial.

On the basis of archeological findings and studies of existing primitive peoples the development of religion is commonly described as a three-stage-processs: animism, polytheism, and monotheism (Park 1989 pp. 569-616).

- Animism: Primitive, nomadic humans, feeling utterly dependent on nature, saw every natural object – tree, rock or stream – as endowed with a spirit. These spirits were thought not only to control the existence of their objects (a tree spirit makes the tree grow and spread its branches; a stream spirit makes the water flow) but indirectly also to influence human life by providing timber, water, etc. Rituals were used to ensure the favour of these spirits and thus to ward off the evil of their loss.
- Polytheism: Gradually, these spirits came to be seen as deities with their own personalities, whom one had to please with gifts, sacrifices, in order to survive. These deities were given names, and were usually associated with forces of nature: storm, rain, and thunder. In addition, tribes commonly adopted a territorial god, such as the Canaanite *Baals* and *Els* in the OT. Gradually, one deity came to be seen as more powerful than the others; this deity is featured as the creator god in ancient creation stories (Van Wolde 1996).
- Monotheism: In the OT we can trace the extended struggle that it took for the people of Israel to advance from polytheism to monotheism (Armstrong 1993). Interestingly, this transition occurred in tiny Israel rather than in the great nations of Egypt, Greece and Rome. During the Exodus the Israelites chose Yahweh, the territorial wilderness god

from Mount Horeb, as their guide and protector, but still only as a tribal god. After many instances of apostasy, related in Judges, Kings and Chronicles, their experience during the Babylonian exile led to the conviction that Yahweh is the universal, omnipresent God, the God of all peoples (Is.49:6), the Creator of everything that is, the eternal and only God (Isa.43:10; 45:5-7,18), to whom even cosmic forces are small and insignificant (Isa.40:12-15,28). Yahweh also came to be experienced as a loving and caring God (Isa.40:11), who seeks a personal relationship with his human creatures and gives them the Law to live by. This posed the question: How can the perfect Yahweh forgive transgressions of his divine Law without compromising his perfect justice? Initially this led to the image of a vengeful god, who ruthlessly punishes the sinner. The prophet Jeremiah predicted that the old covenant of Mount Horeb would be replaced by a new covenant "written upon the heart" (Jer.31:31-34), but he failed to answer the problem that to the ancient mind a valid covenant required the blood of a sacrifice. Other prophets predicted the coming of a Messiah (Mic.5:2-5; Zech.9:9-10), the suffering servant in Second Isaiah (Isa. 42:1-4, 49:1-7; 50:4-9, 52:13-53:12), who will bring reconciliation between Yahweh and his people. This leads me to recognize a fourth stage:

- Trinitarian monotheism: Six centuries after Jeremiah's prophecy the Jewish followers of Jesus of Nazareth (again a small minority) recognized in him the promised Messiah, who through his death on the cross brings reconciliation, a new covenant. Through their experience of his resurrection they came to see him as the incarnate Son of God. The pentecostal experience in Jerusalem led to the awareness of the Holy Spirit as our lasting link with God the Father. The Christian Church was born, grew rapidly and spread over the entire world. During the first four centuries AD the experience of the Apostles was formulated by the Church in the trinitarian monotheistic doctrine of the one God in three persons, Father-creator, Son-redeemer, Spirit-communicator.

Nowadays, many consider the development of religion, sketched here, as a mere human construct. Therefore, it is not surprising that brain scanning techniques have been applied to study cerebral activities during meditation. In a pioneer study, electroencephalography (EEG) was applied to nuns during contemplative prayer (Mallory 1977). Three phases could be distinguished, all of which were accompanied by an increased alpha wave (8-13 Hz): (1) a phase of concentration and inner recollection; (2) a phase of libidinal concentration on Christ as the object of desire; (3) a phase of ecstatic encounter with the object of union. Each phase continued as the next one arrived. The changes in the alpha wave reflected the depth of contemplation reached, as determined by questionnaires subsequently completed by the subjects.

Later EEG studies showed increases in alpha wave, theta wave (4-7 Hz), and in experienced meditators also in beta wave (20-40 Hz) (Newberg et al. 2001). Alpha wave activity is associated with calm and focused attention; theta wave activity with reverie, imagery, and creativity; high beta activity with highly focused concentration. Other experiments, involving PET scanning of meditating Buddhist monks, enabled such effects to be localized. Prefrontal lobe activity (attention and concentration) was increased, while parietal lobe activity (which provides the senses of orientation in space and time) was decreased. The latter effect agrees with the statement of meditators that during meditation their self becomes united with all creation and with the divine.

A few years ago, a claim was made that "the God gene" had been found (Hamer 2004). Participants in a 1000-subject survey were asked to complete the 240-question Temperament and Character Inventory (TCI). One of the traits measured is spirituality or self-transcendence. Nine genes involved in the release of monoamine neurotransmitters such as serotonin (regulating mood experience) were then studied. There was a correlation between the TCI score for spirituality and the gene *VMAT2* (vesicular monoamine transporter). Persons with cytosine at a particular site of this gene had a high TCI score, those in whom this cytosine was replaced by adenine had a low score. The cytosine-type *VMAT2* was therefore considered to be a gene for spirituality. The finding of an increased serotonin level in meditating Buddhist monks suggests that this gene might confer their meditating skill, but its presence was not determined in that study.

To conclude from these studies that religious experience and spirituality are *merely* a function of neurons or genes, as some do, is in my view a form of unwarranted reductionism. I suggest that transcendent thought and spirituality are the result of the *interaction* of divine revelation with the human mind, mediated by the Spirit. Since humans are said to be created in the image of God (Gen.1:26,27), which image was not lost by the fall (Gen.5:1; 9:6), such interaction would seem to be possible.

6. Mind in death and resurrection

What happens to the mind at death, in the light of the expected resurrection on the last day? On biblical grounds I have concluded that a living person is a body-mind unity made alive by God's life-giving Spirit (section 1), and the body-mind unity appears to be confirmed by the neuroscientific findings (section 3). Biologically, we can say that at death the mind stops functioning (as indicated by a flat EEG), and the body begins to decompose. Theologically, we can say that at death God withdraws his life-giving Spirit. About the death of Jesus, John writes: he "gave up his spirit" (Jn.19:30; NRSV), but the Greek text actually has *to pneuma, the* spirit. The synoptic gospels all have: "breathed his last", which can be taken as referring to "spirit" as well as "breath".

At the resurrection a new body-mind unity is made alive by God's spirit, as metaphorically described in Ezek.37:1-10. About the resurrection body we see some glimpses in the accounts of the appearances of the risen Jesus. In some sense the resurrection body must resemble the natural body because Jesus is recognized by the women and the disciples (Mt.28:9, 17; Jn.20:20), although he is not immediately recognized by the men going to Emmaus (Lk.24:16), by Mary Magdalen (Jn.20: 14) and by the disciples at the Sea of Tiberias (Jn.21:4). The most striking difference is the ability of the risen Jesus to walk though a closed door (Mt.28:9; Lk.24:36; Jn.20:19, 26; 21:4) and in a moment to appear, disappear and appear at another site (Mt.28:9; Lk.24:36; Jn.20:19). This suggests that the resurrection body is not bound by the laws of space and time, as Jesus was – like us – during his life on earth before the crucifixion.

But now the question arises: What occurs in the interim period between death and resurrection? The bible is vague on this. In the OT there is *Sheol*, the abode of the dead (Gen.37:35; 42:38; 1Sam.2:6; Job 14:13) and rarely in later and apocryphal writings *Gehenna*, the place of eternal punishment of the dead (Dan.7:10; Enoch 18:11-16, etc.). In the NT *Sheol* does not occur, only *Gehenna* translated as "hell" (Mt.5:22; Mk.9:43; Lk.12:5; Jas.3:6). This is strange, because the idea of eternal punishment following death conflicts with the resurrection of all dead persons on the last day with judgment (Dan.12:2:

> Many of those who sleep in the dust of the earth shall awake, some to everlasting life, and some to shame and everlasting contempt   (Dan. 12:2);
>
> And all were judged according to what they had done   (Rev.20:12-15).

The "near-death experiences" of persons resuscitated after cardiac arrest seem to provide some information (van Lommel *et al.* 2001; van Lommel 2007). Up to 30% of them report that, having entered "the other world", they saw in a flash their entire life as in a 3-D movie, with an overwritten experience of judgement. This would suggest that at death we are not so much judged as that we judge ourselves, in answer to the question: "Am I able and desirous to live in God's presence, thus to believe in him – yes or no?" In the intermediate state there may then be an opportunity for further spiritual growth, until at the last day we shall judge ourselves definitively to either living eternally in God's presence (heaven) or to existing for ever in his absence (hell). In the presence of the light of Christ there will be full disclosure, such that our self-judgment cannot be false. This does not mean "universalism", acceptance of all humans. The Doctrine Commission of the Church of England has expressed a similar view (Doctrine Commission, 1996).

This still leaves the question as to what remains of the person in the interim period. I attempt to answer this question by means of a digital video metaphor. I suggest that during our lifetime a continuous video recording is made of what makes each of us unique, our genome and mind, which is stored in a heavenly computer. At our death and on the last day the entire video is shown to us for our "self-judgement". During the interim period, the video camera continues to record our spiritual growth. On the last day, if we say "Yes!" to God, the video is "printed out" in heavenly quality in a single, three-dimensional hologram, to which God imparts his life-giving Spirit. And there we are, come to life in the resurrection body.

7. Discussion

A survey of the biblical data on the mind reveals the position that humans are a body-mind unity and that there is no real distinction between mind and soul. For this reason, I prefer to eliminate the "soul" concept, considering the mind to encompass both our intellectual and spiritual faculties. This view is supported by current neuroscientific insights. The central nervous system pervades the entire body and is intimately linked with the various organs. The mental functions influence many organic functions, and vice versa (HPA axis). However, we must recognize that neuroscience shows us only the biological substrate of the acting mind, it does not explain the *essence* of the functioning human mind. Thinking it does, constitutes an unwarranted form of reductionism. This is even acknowledged by avowed atheist and sceptic Michael Shermer (2008).

It is regrettable that the patristic and scholastic theologians shifted their thinking to the soul and its moral aspects, thereby neglecting the mind. And this has not changed in the work of modern theologians. Keith Ward wrote a book *In Defence of the Soul* (1998), which is actually an attack on materialism, rather than a discussion of the relation between soul and mind. Philip Rolnick in his *Person, Grace, and God* (2007 pp. 246-247) takes into account the findings of neuroscience, but distinguishes between soul and mind as a *twoness* that is to be unified and that has a higher and a lower level. While not stating this explicitly, Rolnick seems to think that the soul serves religious experiences and the mind intellectual activities. To me this seems to be an unnecessary, even undesirable, compartmentalization of the mind, for which there is neither scriptural warrant nor neuroscientific support.

A striking finding is that in the evolutionary development of the mind we see no clear demarcation between non-human primates and humans, only differences in quality and refinement. This is true for brain microstructure, genome, tool making, mathematical skills, language, and diverse traits (empathy, altruism, self-awareness, cooperation, and learning). Traditional theology assumed that God introduced a soul at the creation of the first human and repeats this at the birth of each human being. Karl Rahner concludes from a theological and scientific study that we may never find the border between animal and human (Rahner 1965). Emil Brunner posits, without much argument, the introduction of the human soul at some point in the hominid line, e.g., when transcendental awareness arose (Brunner 1952 pp. 79-88). However, there is an evolutionary development of religion, which theologically may be seen as a deepening interaction of divine revelation with the human mind. And we are still uncertain where transcendental awareness may have begun. The evolution of religious belief is repeated in broad outline during individual human development (Fetz *et al.* 2001). *Hominization* is thus a continuous process of physical and spiritual development in human evolution as well as in the growth of each individual.

I briefly mention in section 5 that synaptic transmission appears to be subject to "chaos events". This is a very important finding, because it means that our mental processes are open to God's influencing of chaos events through the action of the Spirit, as I have explained elsewhere in a general sense (Bonting 2005 pp. 115-124).

Rather than the insertion of a soul, I conclude that God brings life to the body-mind unity by means of the life-giving activity of the Spirit (Bonting 2006). At death God withdraws this life-giving Spirit, the mind stops functioning and the body begins to decompose. In the biblical account there is no surviving soul, in contrast to Greek philosophy and much popular belief. In the resurrection on the last day, we shall appear in a transformed body-mind unity. From the scarce biblical data I have constructed a metaphorical idea for the interim period between death and resurrection. It is regrettable that insufficient attention is given to creation theology in the endless debate between resurrection sceptics and defenders of the resurrection of Christ and, derived from this, the resurrection of the dead (McGrath 2001 pp. 397-405). Belief in creation implies completion of the creation process, which necessarily entails resurrection of the Son and of all the dead.

References

Amodio, D.M. *et al.* 2007. Neurocognitive correlates of liberalism and conservatism, *Nature Neuroscience* 10, 1246-1247.
Armstrong, K. 1993. *A History of God: From Abraham to the Present, The 4000-Year Quest for God*. London: Heinemann.
Balter, M. 2007. Brain evolution goes micro, *Science* 315, 1208-1211.
Bonting, S.L. 2005. *Creation and Double Chaos*. Minneapolis: Fortress Press.
—. 2006. Spirit and Creation, *Zygon: Journal of Religion & Science*, 41 (September), 709-722.
Britannica, The New Encyclopaedia, 1989, 15th ed., vol. 8, article Mind. Emotion and religious experience added here to the Encyclopaedia list.
Brunner, E. 1952. *The Christian Doctrine of Creation and Redemption. Dogmatics* vol. II, London: Lutterworth Press, London, 1952.
Cheney, D.L. & Seyfarth, R.M. 2007. *Baboon Metaphysics; the Evolution of a Social Mind*. Chicago: University Press.
Cohen, J. 2007a. The world through a chimp's eye, *Science* 316, 44-45.
—. 2007b. Relative differences: The myth of 1%, *Science* 316, 1836.
Dantzer, R. *et al.* 2008. From inflammation to sickness and depression: when the immune system subjugates the brain, *Nature Reviews Neuroscience* 9, 46-56.
Dentan, R.C. 1962. Mind, in *The Interpreter's Dictionary of the Bible,* vol 3. Nashville: Abingdon.
Doctrine Commission, Church of England, 1996. *The Mystery of Salvation, The Story of God's Gift*. London: Church House Publishing.
Dunbar, R.I.M. & Shultz, S. 2007. Evolution in the social brain, *Science* 317, 1344-1347.
Dupue, B.E. *et al.*, 2007. Prefrontal regions orchestrate suppressions of emotional memories via a two-phase process, *Science* 317, 215-219.
Fetz, R.L. *et al.* 2001. *Weltbildentwicklung und Schöpfungsverständnis. Eine struktur genetische Untersuchung bei Kindern und Jugendlichen*, Stuttgart: Kohlhammer.
Frank, M.J. *et al.* 2007. Hold your horses: Impulsivity, deep brain stimulation, and medication in Parkinsonism, *Science* 318, 1309-1312.

Ghazanfar, A.A. 2008. Language evolution: neural differences that make a difference, *Nature Neuroscience* 11, 382-384.
Gunter, C. 2005. The chimpanzee genome, *Nature* 437, 47-64.
Hamer, D.H. 2004. *The God Gene: How Faith Is Hardwired Into Our Genes.* New York: Doubleday.
Herrmann, E. *et al.* 2007. Humans have evolved specialized skills of social cognition, *Science* 317, 1360-1366.
Horgan, T.E. 1994. Physicalism in *A Companion to the Philosophy of Mind,* S. Guttenplan (ed.). Oxford: Blackwell.
Kauer, J.A. & Malenka, R.C. 2007. Synaptic plasticity and addiction, *Nature Reviews Neuroscience* 8, 844-858.
Knight, R.T. 2007. Neural networks debunk phrenology, *Science* 316, 1578-1579.
Laureys, S. 2007. Eyes open, brain shut, *Scientific American* 296 (May), 66-71.
Ledford, H. 2007. Monkeys add up like we do, *Nature* online (18 December)
Levina, A. *et al.,* 2007. Dynamical synapses causing self-organized criticality in neural networks, *Nature Physics* online, 18 November.
Libet, B. 1985. Unconscious cerebral initiative and the role of conscious will in voluntary action, *Behavioral and Brain Sciences* 8, 529-566.
McGrath, A.E. 2001. *Christian Theology* (edn 3). Oxford: Blackwell.
McNab, F. & Klingberg, T. 2007. Prefrontal cortex and basal ganglia control access to working memory, *Nature Neuroscience* 11, 103-107.
Miller, G. 2007. A surprising connection between memory and imagination, *Science* 315, 312.
—. 2008. Mirror neurons may help songbirds stay in tune, *Science* 319 (18 January), 269.
Minkel, J.R. 2008. Food for symbolic thought, *Scientific American* 298 (January) 16 (see also 86-87).
Mithen, S. 2005. *The Singing Neanderthals,* London: Weidenfeld & Nicholson.
Newberg, A. *et al.* 2001. *Why God Won't Go Away: Brain Science and the Biology of Belief.* New York: Ballantine.
Nicolelis, M. & Ribeiro, S. 2006. Seeking the Neural Code, *Scientific American* 295 (December), 48-55.
Nitz, D & Cowen, S. 2008. Crossing borders: sleep reactivation as a window on cell assembly, *Nature Neuroscience* 11, 126-128.
Park, G.K. (ed.) 1989. Systems of religious and spiritual belief, *Encyclopaedia Britannica,* 15th ed., Chicago: Encyclopaedia Britannica Inc., vol. 26.

Pennisi, E. 2007. Nonhuman primates demonstrate humanlike reasoning, *Science* 317, 1308.
Porteus, N.W. 1962. Soul, in *The Interpreter's Dictionary of the Bible*, vol 4. Abingdon, Nashville.
Rahner, K. 1965. *Hominisation, The Evolutionary Origin of Man as a Theological Problem.*, Freiburg: Herder/London: Burns & Oates.
Rizzolatti, G. & Craighero, L. 2004. The mirror-neuron system, *Annual Review of Neuroscience* 27, 169-192.
Robinson, H.W. 1913. *The Christian Doctrine of Man* (edn 2). Edinburgh: T & T Clark.
Rolnick, P.A. 2007. *Person, Grace, and God*, Grand Rapids, MI: Eerdmans.
Schacter, D.L. & Addis D.R. 2007. The optimistic brain, *Nature Neuroscience* 10, 1345-1347.
Scharfman, H.E. & Hen, R. 2007. Is more neurogenesis always better?, *Science* 315, 336-338.
Segerstrom, S.C. & Miller, G.E. 2004. Psychological stress and the human immune system: A meta-analysis of 30 years of inquiry, *Psychological Bulletin*, 130, 601-630.
Shermer, M. 2008. A new Phrenology?, *Scientific American* 298 (May), 25-26.
Silk, J.B. *et al.*, 2005. Chimpanzees are indifferent to the welfare of unrelated group members, *Nature* 437, 1357-1359.
Silk, J.B. 2007. Social components of fitness in primate groups, *Science* 317, 1347-1351.
van Lommel. P. *et al.* 2001. Near Death Experience in survivors of cardiac arrest: Prospective study in the Netherlands, *The Lancet*, 358, 2039-2042.
—. 2007. *Eindeloos bewustzijn* [Endless consciousness], Ten Have: Kampen.
van Wolde, E. 1996. *Stories of the Beginning*. London: SCM Press.
Mallory, M.M. 1977. *Christian Mysticism: Transcending Techniques*, Assen/Amsterdam: Van Gorcum.
Ward, K. 1998. *In Defence of the Soul.* Oxford: Oneworld.
Whiten, A. *et al.*, 2005. Conformity to cultural norms of tool use in chimpanzees, *Nature*, 437, 737-740.
Whitfield, J. 2008. Paging Dr. Doolittle, *Scientific American* 298 (January), 13-14.
Wikipedia, The Online Encyclopaedia, 2008. Article: Philosophy of Mind, http://en.wikipedia.org/wiki/Philosophy_of_mind

Wynn, T. & Coolidge, F.L. 2008. A stone-age meeting of minds, *American Scientist* 96, 44-51.

Zimmer, C. 2005. The neurobiology of the self, *Scientific American* 293 (November), 64-71.

INDEX

In this index, only nouns are listed, although some of the text words may be adjectives, e.g. 'Darwinian' and 'Darwinism' (text) both appear here under 'Darwin'. There are also a few occasions on which the listed page explores the named idea without using the specific word.

Names of people are indexed only if they carry a total of at least six-eight lines of associated (though not always immediately associated) text. Names of contributors to this volume are indexed at pages where they are referred to in the text; for the pages carrying their own contributions, see the Table of Contents (pp. v-vii).

adaptation 35, 43-4, 78, 94-5, 97, 142, 166-7
adaptationist theories 43-7, 62, 203
African origins 14-8, 21-2, 24, 136-7, 149
agriculture 22
Alzheimer's disease 124
analogies, exaggeration of 58
animism 12, 180, 228
Anthropic Principle 158
anthropomorphism 39
Aquinas, Thomas 90, 113, 140-1, 171-3, 217
archeological record, identifying religiosity in 13
artefacts, symbolic 17-9
atheism, cognitive effort of 26
Augustine (of Hippo) 47, 54, 84, 134, 217
Aurignacian culture 22
autopoietic systems 191-2
awareness
    animal 108, 226
    levels of 46-7, 77, 98-9
axon 220

Barrett, Justin 33-4, 75, 86-7, 96 *footnote*
Barth, Karl 134, 144, 146, 150
beauty in mathematics 185, 187
behaviour prediction 16
behaviourism 219
Bekoff, Mark 38-9
Berekhat Ram "figurine" 17
Berkeley, George 218
Berry, Thomas 195-6
Biblical criticism 55, 69
Bilsingsleben bones 18
biological diversity 190-1
biology, as starting point 74-6
bipedalism 15, 150, 227
Blombos cave 22-4, 137
body-mind unity 3, 6-7, 214, 231-3
Boff, Clodovis 133
Bonting, Sjoerd 6
Boyer, Pascal 12, 25, 35, 65, 75, 86-7, 90
brain
    regions of 123, 220-5
    size of 17, 135, 165
Brunner, Emil 234
burials
    Mount Carmel 18, 23-4

Neanderthal 19, 228
  Shandidar 14, 150
by-product theories 44

Caduceus, The 206-7
Calvin, John 218
Campbell, Donald T 76, 81-3, 88, 95-6, 98
category mistake 106
causation
  mutual 122, 124
  top-down 116, 120, 168
cave art 23-4, 130-1, 137
chimpanzee/higher apes 225-7
  religious awareness in? 2, 38-9
  theory of mind 108
  tool-use by 15, 142
Choong, Ron 6
Christ Jesus 4, 60-2, 133, 138-140, 144, 147, 150, 152, 158-9, 164, 171-2, 201, 207, 229, 231, 234
Clayton, Philip 122-126
Clement (of Alexandria) 217
cognitive
  domains 21, 24-6, 39, 63, 65, 100
  fluidity 2, 5, 24-7, 33-5, 39-41, 63-4, 100, 136, 150-1, 164-5
  psychology 64, 219
  science 65, 115
communication 136, 167
  religious 49
  verbal 16
comprehensibility, miracle of 184
concepts
  physical/objective 78, 84-6
  metaphysical 84, 86, 94, 170, 174
  social 76
consciousness 97-100, 105-7, 123, 136, 159, 164-5, 194, 201-4, 208-11, 214, 216, 224
  orders of 60, 62, 71
constructivism 187
contemplative prayer 229

contingency 138, 140-1, 143-4, 149, 155-60, 171-4
  and necessity 4, 156-9
Copernicus, Nicolaus 190
counter-adaptive traits 43
counter-intuitive concepts 25-6, 49, 63-65, 134
creation, ongoing 55-7, 69-70, 163-4, 171-2
creativity 185-6
culture 2, 34, 37, 43-5, 50-1, 55-6, 98, 100, 166-7, 179-87

Darwin(ism) 12, 14, 34, 53-66, 69-71, 76-90, 93, 96-7, 100, 102, 104, 108, 113, 131, 166, 191, 194, 201
Dawkins, Richard 12, 59, 72, 90, 96, 109, 209-11
Deacon, Terrence 36, 36 *footnote*, 49, 135, 137-8
Deane-Drummond, Celia 2-3.
death 19, 35, 87, 105, 113, 131-3, 151-2, 195, 215-7, 228-34
deist view 57,.69-71
dendrite 220
Dennett, Daniel 43, 108-12, 120, 132
Descartes, René 75, 107-110, 183, 218
dignity, human 57-8
direction in evolution 55-7, 65, 69, 71, 131, 149, 162, 172-3, 210.
  See also progress
Dow, James 44
Drees, Willem B 110-1
dualism 3, 105, 107-114, 116-7, 119-22, 126, 167-8, 174, 218
Duns Scotus 217

electroencephalography (EEG) 221
emergence 116-7, 120, 122-7, 131, 135-6, 140-3, 148-150, 162-3, 168, 193-6, 204, 209-11
emotion and religion 48
empiricism 79, 81, 106, 109, 183

enablement v. causation 34-5
evil 56, 58, 60-1, 65, 70-2, 113-4, 164, 204
evolution 12, 14-5, 24, 26, 33-5, 42-51, 53-66, 68-72, 85, 87-90, 94-100, 104, 108, 111, 130-5, 138-52, 155-6, 158-67, 171-5, 179, 185-6, 191-3, 195-6, 201, 204, 206-11, 223, 225-8
  of complexity 56, 162-3, 168
  of mind 15-6, 104, 116, 138, 142, 175, 186, 195-6, 214, 225, 233
  of morality 59, 72, 201, 204
  of religion 1-2, 42-51,.62-5, 72, 214, 228, 233
evolutionary epistemology 75-90, 93, 95-8, 100-1, 166-7, 174-5
evolutionary psychology 2, 4, 32-3, 96, 114, 167
  objections to 34-6, 39-40, 100

folk psychology 106-16
foundationalism/non- 75, 96
free will 105, 199, 201-4, 217-8, 224
Freeman, Anthony
functional MRI (fMRI) 221
functionalism 219
fundamentalism 54, 69, 90

Genesis, Book of 54, 60, 139, 149, 164, 180, 203-4
genetic fallacy 58
ghost in machine 107-8, 119, 127
God gene 230
God a mathematician 184
God of the gaps 3, 7,
God, working through natural forces 33
Gould, Stephen J 34, 131, 147
graves, *see* "burials"
Gregory of Nissa 217
guilt 200, 204

healing 206-8, 210

Hegel, Georg WF 148, 159
Hitchcock, Gavin 5
Hohenstein Stadel "lion man" 23
hominins, Plio-Pleistocene 14
hominization 234
Homo, early species of 15-7, 135-6, 150, 165, 191, 227
HPA axis 225
Hume, David 82, 81, 117, 123, 157, 159

Idealism 81,117, 218
identity theory 219
Iannaccone. Lawrence 47
imagistic v. doctrinal religiosity 12-3, 23-4, 32, 50
imago Dei 37-8, 71, 111, 117, 120, 125-7, 149-51, 172, 204
inborn propensity 79-82
incarnation 60-1, 71, 111, 114, 127, 134, 138-40, 145-7, 149, 152, 158, 163-4, 171-2, 211
innate ideas 79-81, 96-7
intentionality of events 23
interim period 231

John, St 4, 139, 144, 146, 152 *footnote*, 216, 231

Kant, Immanuel 64, 79-81, 97
Kauffman, Stuart 5, 56, 162-3
Knight, Roger 4

language 21, 24-5, 60, 94-5, 97-102, 106-8, 124, 136-9, 150, 165, 169, 175, 194, 202, 223, 225-8, 233
  2- v. 3-dimensional 36-8, 98
  metaphorical 64
Law, Jeremy 4
leb 215
levels of description 7
Libet,Benjamin 202-4, 224
life-giving spirit 217
living species, number of 191

Lockwood, Michael 122-3
Lorenz, Konrad 76-7, 79-82, 85, 94-5, 97
Lovelock, James 189, 191-4
Luther, Martin 218

magnetic resonance imaging (MRI) 221
mathematics 5, 179-187, 226, 233
meaning 11, 25, 36, 46, 49, 64, 84-5, 94-5, 100-1, 111, 132, 136, 144, 155-7, 208, 211
Melanchton, Philip 218
memory 221-4, 226, 228
mentality
  domain-specific 20
  Neanderthal 20-1
metaphor 24, 64, 161, 165, 169-71, 174, 202, 204
metaphysical entities 105, 109, 113, 115
metaphysical theology 83-5, 90, 94-5, 101-2, 113, 117, 125, 170-1, 174
mirror neurons 223
Mithen, Steven 1-5, 31-3, 35-6, 39-40, 43, 63, 100, 131, 135-7, 150, 161, 164-5, 169, 171, 202
monism 4, 110, 114, 117, 218
monotheism 228
monotheism, Trinitarian 229
Moore, Aubrey 2-3, 53, 57, 69
morality 22, 37-8, 58-9, 66, 72, 109, 190, 194-5
Morris, Simon Conway 56, 70, 133, 142-3, 150, 158, 162, 172-3, 209
Munz, Peter 36-8, 76, 80-2, 87, 97-100, 102

naturalism 88, 110, 112-3, 115, 117, 203
Neanderthals 15, 17, 20-1, 32, 38-9, 135-7, 152, 165, 227-8
near-death experiences 231
nephesh 215

neshama 215
nested hierarchy 121
neural network 222
neurological view of religion 45
neuron 220
neurobiology/neuroscience 3, 83, 97, 102, 116, 120, 120-3, 125, 200-2, 214, 218-20, 232-3
neurotransmitter 221
Newberg, AB & d'Aquili, EG 122-3, 230
Nicolas of Cusa 126
nolition 199, 203
nous 216

Origen 217
Ovieod, Lluis 2, 5

paleo-anthropology 1, 4
panentheism, 113, 125-7, 149
Paul, Roger 5
Paul, St 89, 114, 208, 216
Peacocke, Arthur 35, 56, 60, 70, 88, 126, 138, 143, 207
personal identity 105
personality 211, 216, 222
physicalism 110, 112-5, 117-8, 122, 218-9
  reductive 218
  non-reductive 117, 122, 219
plasticity 223-4
Plato 5, 74, 79-80, 114, 183-4, 187, 217
pneuma 216
polytheism 228
Popper, Karl 3, 76, 82, 85, 94-98, 100-1, 103, 116, 143
positivism 3, 96-98, 109-10
positron emission tomography (PET) 222
problem-solving animals 178
progress 11, 54-6, 61-2, 81, 162, 172-3, 210
psyche 216
Putnam, Hilary 219

Rahner, Karl 60, 140, 233
rationality and religion 47-8, 183-4
readiness potential 203, 224
reality, conceiving/modeling of
    178, 183-7, 194-5
reciprocities, mysterious 183
redemption: *see* salvation
reductionism 43, 58-9, 120, 122,
    169, 174, 202, 230, 232
religious experience 46-7, 110, 114
religiosity/religious mind 10-3,
    24, 31-2, 138
    Absence before *H. sapiens*
    20
    imagist v. doctrinal mode 12,
    32, 169
resurrection 113, 159, 201, 214,
    229-34
Rolnick, Philip 232
Rose, Stephen 34-6
ruach 215
Runehov, Anne 4
Ryle, Gilbert 106-8

salvation/redemption 71-2, 134,
    139, 152, 156, 163-4, 200-1,
    217-8
selection
    group 44, 81
    natural 11, 33-8, 43-5, 49-51,
        70, 76-90, 93, 102, 131,
        162, 165-7, 169-70, 175,
        186, 191
self
    concept of 110-13, 116, 120-7
    link with God-concept 112-4,
        116, 120, 167
    threefold 121, 123-7
sense organs 76-7, 94, 166
self-awareness 47, 62, 98-9, 106,
    123, 131, 136, 138, 143, 168,
    175, 181, 211, 224, 233
self-organisation 45, 191
sin 199, 202-4, 217
Smart, J.J.C 219

social grooming 16
soul 80, 84, 87, 105, 111, 113, 167-
    8, 214-9, 232-3
Spinoza, Baruch 218
spirituality 10, 90, 112, 230
Spurway, Neil 3, 93-9, 101-2
Stanesby, Derek 3-4
supernatural beings 11-2, 24-6, 35,
    84, 86
supervenience 4, 122, 168
survival 3, 137-8, 165-6, 175, 180,
    201, 210
    benefits of religion to 43-7, 49-
        50, 62, 155-6, 169-70, 203
    of concepts 3, 78-9, 81-2, 85-7,
        94
    struggle for 61, 70, 77-9, 137,
        166, 175
Swayne, Jeremy 6
Swiss Army knife 33, 40
symbolic thought 22, 48-9, 133-7,
    139, 151
symbolism and religion 48-9, 131,
    169, 174
synapse 204, 220-2

Taliaferro, Charles 116-7
Tattersall, Ian 89, 137, 140, 143
Tertullian 217
theodicy 131-2
thermodynamics of religion 45
tool-use
    by chimpanzees 15, 226-7, 233
transcranial magnetic stimulation
    (TMS) 222
Trigg, Roger 3, 96 *footnote*
Trinity 133-5, 144-7, 149, 151, 173

Umwelt 77, 166, 170
unexpected liaisons 182
universal acid 43

Vane-Wright, Richard 6
van Huyssteeen, Wentzel 37-8, 87-
    9, 125, 134-5, 137-8, 149
via negativa 90, 174-5

Vollmer, Gerhard 76, 79, 82, 85, 94

Ward, Keith 232
Watts, Fraser 2-4, 68, 70, 72, 172, 201-2, 208
Weber, Max 47-8
Whitehouse, Harvey 12-4, 23, 32, 50

wholeness 207-10
Wilson, David S. 44, 75, 87
Wilson, Edward O. 34 *footnote*, 59
Wittgenstein, Ludwig 99-100, 106, 194, 219
Word 114, 139-40, 146-9, 151-2, 158-9, 171-2, 194